Kirstenbosch Gar

GROW
CLIVIAS

A GUIDE TO THE SPECIES, SELECTED HYBRIDS, CULTIVATION AND PROPAGATION OF THE GENUS *CLIVIA*

Text by Graham Duncan
Photographs by Graham Duncan unless otherwise acknowledged

Clivia miniata 'Kirstenbosch Supreme'

Clivia Minirabilescent Group bred by John Winter at Kirstenbosch (see pages 110, 111)

John Winter admiring a population of *Clivia gardenii* var. *citrina*, KwaZulu-Natal

Dedicated to

John Winter

In recognition of his tremendous contribution to the cultivation, propagation and habitat collections of clivias, which he has made with great enthusiasm and dedication.

John Winter has had a long and successful career promoting indigenous South African plants in general, and *Clivia* in particular. He was Curator of Kirstenbosch from 1978 to 1998, and apart from all his other tasks, was also personally responsible for collecting and breeding a number of genera, including *Disa, Erica, Protea, Strelitzia* and *Clivia*. An *Erica* and a *Leucospermum* have been named after him, and he bred the *Disa* hybrid 'Kirstenbosch Pride'. In 1998 he was appointed as Deputy Director of the then National Botanical Institute, and was responsible for the management of all the National Botanical Gardens until his retirement in 2001. In 2001/2 he became the first Chairman of the *Clivia* Society, and has been Chairman of the Cape *Clivia* Club since its inception.

John Winter was aware that very few of the *Clivia* clones at Kirstenbosch had been collected in habitat, and that there were very few specimens in any of the herbaria in South Africa. Therefore, in 1995, he went to the Eastern Cape to locate *Clivia* in the wild, and collect them for the habitat collection at Kirstenbosch, now the most comprehensive in the country. As the plants have multiplied, he has distributed them among the other National Botanical Gardens so that they can be enjoyed by all. He has returned to the Eastern Cape and extended his searches into KwaZulu-Natal, Swaziland and Mpumalanga; these searches have established the boundaries of all the summer rainfall *Clivia* species, which he updates as new colonies are discovered. When the SANBI identified *Clivia mirabilis* as a new species, it agreed to germinate and grow the first seeds collected by the Northern Cape Department of Nature Conservation; John was entrusted with implementing this agreement and marketing the first seedlings around the world. As if that were not enough, he has continued his own successful breeding of new and prize-winning *Clivia* hybrids.

Sean Chubb, Chairperson, KwaZulu-Natal Clivia Club
Clivia Society Newsletter, winter 2004

Clivia miniata in the Camphor Avenue at Kirstenbosch

CONTENTS

A brief history 1
General information 21
   Introduction 21
   Growth and flowering cycles 26
   Stems and foliage 27
   Inflorescence and flowers 30
   Fruits and seeds 36
   Pollination 40
   Distribution and habitat 42
   Medicinal uses and conservation 49
Taxonomy 55
   Generic description 55
   Key to the species 56
   Clivia miniata 58
   Clivia caulescens 72
   Clivia mirabilis 76
   Clivia nobilis 82
   Clivia robusta 88
   Clivia gardenii 94
Hybrid, variegated & novelty clivias 101
Cultivation 133
   Aspect and climate 133
   Hardiness 135
   Clivias in the garden 135
   Container subjects 139
   Growing medium 139
   Watering 141
   Feeding 144
Propagation 147
   Seed 147
   Pollinating clivias 153
   Division 154
   Tissue culture 155
Pests and diseases 157
References and further reading 169
Glossary 177
Index 180
Useful addresses 185

Below: *Clivia miniata* Pat's Gold strain, bred by the author at Kirstenbosch (see page 117)

Opposite: *Clivia nobilis* (pinkish-yellow form) from the Eastern Cape (see page 82)

A BRIEF HISTORY

The English naturalist, traveller, artist and author, William J. Burchell (1781–1863), was the first person to make a scientific collection of a *Clivia* species (*C. nobilis*) in the wild, which he did near Port Alfred in the Bathurst district of the Eastern Cape, in September 1813 (Baker 1896, Hutchinson 1946). In the latter publication, *A botanist in southern Africa,* the distinguished English horticulturist John Hutchinson (1884–1972) lamented the tendency of gardeners in South Africa at that time to cultivate exotic species, despite the wealth of the indigenous flora. Evidently, the same observation had been made by Burchell more than a century before, for Hutchinson includes a quoted passage from Burchell's *Travels in the interior of southern Africa* (1822) under the title *The travels of William J. Burchell*.

'It may naturally be supposed, that, in a country abounding with the most beautiful flowers and plants, the gardens of the inhabitants contain an great number of its choicest productions; but such is the perverse nature of man's judgment, that whatever is distant, scarce, and difficult to be obtained, is always preferred to that which is within reach, and is abundant, or may be procured with ease, however

Hand-coloured lithograph of *Clivia nobilis* by M. Hart, plate 1182 from *Edwards's Botanical Register* (1828)

Reproduced by courtesy of the Bolus Herbarium Library, University of Cape Town

beautiful it may be. The common garden flowers of Europe are here highly valued; and those who wished to show me their taste in horticulture, felt a pride in exhibiting carnations, hollyhocks, balsamines, tulips, and hyacinths, while they viewed all the elegant productions of their hills as mere weeds.'

While great strides have been taken in the last few decades to promote the cultivation of South African indigenous flora, the lamentable situation described by Burchell more than 150 years ago, remains not greatly improved at present. Had he been alive today, however, he might have taken comfort in the knowledge that the genus *Clivia* has come to command the attention of virtually every gardener in the country, albeit only within the past decade. While Burchell was the first to discover the plant, it was in fact from plants collected by Bowie, possibly in 1822, that the species was named and described by William Hooker (1785–1865) as growing 'on shaded spots, near Quagga flats, and more common in the Albany tracts, near the Great Fish River' (Hooker 1828). The intrepid Kew gardener and botanical collector, James Bowie, arrived in Table Bay from Brazil on 1st November 1816, and until 1818, he collected plants in the vicinity of Cape Town. It was in 1822, during the course of his fourth long plant-collecting journey, which took him to the eastern, southern and western parts of the Cape, that he gathered plants of *Clivia nobilis* in the same area of the Great Fish River as Burchell had. In 1823 he brought them to Kew Gardens, and Syon House, residence of the Duke and Duchess of Northumberland in England, just over the Thames from Kew. The plant first flowered in the conservatory at Syon House, and using this material, the celebrated English botanist and horticulturist John Lindley (1799–1865) who later became secretary of the Horticultural Society in England in 1858, described *Clivia nobilis* in *Edwards's Botanical Register* in October 1828, naming it after the noble Lady Charlotte Florentia Clive, Duchess of Northumberland (1787–1866), who had been cultivating many of Bowie's plants at Syon House.

'This noble plant is supposed to have been one of the discoveries of Mr Bowie at the Cape of Good Hope, from some of the inner districts of which colony it was probably procured. The plant from which our drawing was made, flowered for the second time in July last, in the princely Garden of His Grace the Duke of Northumberland, at Syon House, and was communicated to us by Mr Forrest, to whom we are indebted for several observations upon its habit and characters. We have named this plant in compliment to Her Grace the Duchess of Northumberland, to whom we are greatly indebted for an opportunity of publishing it. Such a compliment has long been due to the noble family of Clive, and we are proud in having the honour of being the first to pay it.' (Lindley 1828).

By a startling coincidence, the English botanist and Director of the Royal Botanic Gardens at Kew, Sir William Jackson Hooker, working independently of Lindley, published a different genus on the same day in October 1828, in volume 55 of *Curtis's Botanical Magazine*, based on the same plant. Hooker named his plant *Imatophyllum aitoni*, in honour of William Townsend Aiton, who was James Bowie's patron. He had intended to name his genus *Imantophyllum*, but in error the 'n' had been omitted, a mistake that was later corrected by him. Speculation at that time regarding the origin of the plant described

3

Hand-coloured lithograph of *Clivia miniata* by James Andrews (c. 1801–1876), a prolific London artist, who contributed many plates to the *Floral Magazine*

Reproduced by courtesy of Dr John Rourke

by Lindley suggested that it may have been 'surreptitiously obtained from Kew', but this was shown to be probably untrue following an investigative account published in *Clivia* 5 (Van der Linde 2003b). Although *Clivia nobilis* was described from material collected by James Bowie that had been the first to flower in England, the plant had been cultivated at least six years previously by the Rev. William Herbert (1778–1847), who had been unable to coax his plant into flower as a result of having placed it in an excessively heated greenhouse. Because both names were published simultaneously, one of them had to give way; and for reasons not immediately apparent, *Clivia* became more widely accepted. Two years later, the German botanists J.A. Roemer and J.H. Schultes formally relegated *Imantophyllum aitonii* to synonymy in *Systema vegetabilium* (Roemer & Schultes 1830).

One of South Africa's showiest amaryllids, the trumpet-flowered *Clivia miniata*, was discovered in KwaZulu-Natal in the early 1850s (see page 58). It was introduced into horticulture in England in 1854 by the brothers James and Thomas Backhouse of Backhouse Nurseries in York (having been received from Andrew Steedman of KwaZulu-Natal), and has been in cultivation for well over one and a half centuries. It was first described by John Lindley as a doubtful species of *Vallota*, as *Vallota ? miniata* in The Gardener's Chronicle (Lindley 1854), commemorating the French botanist Pierre Vallot (1594–1671), because of the resemblance the flowers have to *Vallota speciosa,* the 'George lily', 'Knysna lily' or 'Scarborough lily' (now *Cyrtanthus elatus*). The type specimen, a single pressed flower from a plant belonging to Messrs Lee, is housed in the Lindley Herbarium at the University of Cambridge. Another description of the plant followed later that year in *Curtis's Botanical Magazine,* by W.J. Hooker, who transferred it from *Vallota* to a doubtful species of *Imantophyllum*, as *Imantophyllum ? miniatum* (Hooker 1854). A decade was to pass, however, before the plant was finally placed in *Clivia* by the German botanist E.A. von Regel (1815–1892) as *C. miniata*, published in the gardening magazine *Gartenflora* in Berlin in 1864.

Not surprisingly, *Clivia miniata* aroused the interest of horticulturists and breeders almost immediately after its discovery, and many striking intraspecific hybrids were subsequently raised in England, Belgium, Germany and other countries. The plant was probably introduced into Japan in the late 1850s, and into China in the 1870s, at which time they were grown by the more affluent. Despite their shorter history of cultivation, they are today much more popular in China than in Japan. The first plants grown in China were very different to the ones commonly grown there today; they had long, narrow leaves arranged in loose fans compared with the stiffly arranged, very short- and broad-leafed forms that have been bred for use as indoor container plants. For the Chinese, the major attraction of variegated clivias was the manner in which the leaves imparted an impression of golden silk when placed in bright light.

During the Victorian era in the late nineteenth century, *Clivia miniata* cultivation as an indoor pot plant became the rage in the United Kingdom, as well as in Belgium, Germany and The Netherlands, as the minimally heated home environment suited it very well. Although its popularity decreased in these countries in the early 1960s due mainly to the introduction of central heating, a thriving *Clivia* industry still exists in Belgium today. It is the largest supplier in

Hand-coloured lithograph of *Clivia gardenii* by W.H. Fitch (1817–1892), used by William Hooker in volume 12 (Series 3) of *Curtis's Botanical Magazine* in 1856
Reproduced by courtesy of the Bolus Herbarium Library, University of Cape Town

Europe, annually producing many hundreds of thousands of compact, flowering pot plants with broad, short leaves. Great attention is paid to obtaining early-flowering plants just before Christmas, with inflorescences that extend well above the foliage. Shortly after its introduction in Europe, enterprising growers got busy 'improving' leaf shape and flower colour and shape in *C. miniata*. Through careful selection and hybridization between forms of this species, the Belgian growers, and German growers to a lesser extent, began to develop plants with broader, shorter leaves with rounded tips and markedly upright, tulip-shaped, dark red flowers that became well known as the Belgian Hybrids. Their achievements were taken a few steps further by Japanese growers who produced much smaller, broad-leafed individuals, well known as the 'Daruma' range of strains. Whereas in western culture emphasis has traditionally been placed on flower colour and inflorescence shape, and to a much lesser extent, foliage, in the Far East attention is focused primarily on the leaves. The desired plant in Japan is a compact one with broad foliage, with much emphasis placed on a strict distichous habit (leaves symmetrically arranged in two opposite rows) with the leaves having rounded tips and curved distinctly downwards.

The discovery of a third *Clivia* species occurred in the mid 1850s, shortly after that of *C. miniata*, when Major Robert Jones Garden collected a different pendulous-flowered species in KwaZulu-Natal, material of which was sent to Kew (in a Wardian case supplied by Dr William Stanger, the colonial surveyor-general) together with an interesting collection of other plants, in 1853 (see page 94). Distinguished from all other *Clivia* species except *C. robusta* in that it blooms in autumn and winter, *C. gardenii* is a rather variable taxon across its distribution range, having ornamental, long, distinctly curved, tubular flowers in shades of pale to dark orange and orange-red, or sometimes yellow, with bright green tips and, in most forms, complemented by a well exserted stigma and stamens. Major Garden, a British soldier and avid naturalist, was born in about 1821, and was stationed at Pietermaritzburg in 1848 while serving with the 45th Sherwood Foresters Regiment that was garrisoned there between 1843 and 1859 (McCracken & McCracken 1990). He departed from Durban for England in late 1853, and his *Clivia* was described three years later as *C. gardeni* by W.J. Hooker in *Curtis's Botanical Magazine*, the specific name mistakenly with a single 'i' (Hooker 1856).

The discovery of the first yellow form of *Clivia miniata* in about 1888 in Zululand in KwaZulu-Natal provided yet another sought-after floral prize from South Africa for the attention of gardeners and breeders in England, and subsequently in many other countries (see page 64).

The first published record of a yellow form of *Clivia miniata* from the wild was in 1899 when the English horticulturist, William Watson (1858–1925), included a note about it in the section 'Kew Notes,' in volume 25 of *The Gardener's Chronicle* dated April 15 1899, noted '*Cliveia miniata citrina*. This well-marked, beautiful variety is now in flower in the T- range at Kew. It is said to have been collected wild in Zululand by Captain Mansell, and first flowered in the garden of Mrs Powys Rogers at Burngoose, Perranwell, Cornwall, in April 1897, when flowers of it, and subsequently a little plant, were sent to Kew. An example of the same variety had, however, already been added to the Kew collection by the Rev. W.H. Bowden of Bow, North Devon, who sent it along with some other plants which had been collected in Zululand, and it is this plant which is now in flower. It resembles a good form of the type in every particular except colour, in which it differs widely from all the forms hitherto raised in gardens, and popularly known as Imantophyllums. These are all more or less of a reddish orange colour, but the variety *citrina* is coloured a clear pale cream with a faint tinge of orange at the base of the segments. This variety ought to prove valuable to breeders of Cliveias, whose efforts so far have produced exceptionally little colour variation in the seedlings raised' (Watson 1899a). He followed this with a more detailed description of the plant later that year in volume 56 of *The Garden*, accompanied by a watercolour plate by H.G. Moon of the plant in Mrs Powys Rogers' garden (Watson 1899b). Watson joined the Royal Botanic Gardens, Kew in 1879, where he became Curator in 1901 up until his retirement in 1922. He was a prolific author of horticultural articles (especially in *The Gardener's Chronicle*) and books, and was an authority on indoor plants (John David, personal communication). He visited South Africa during February-March 1887, landing at Port Elizabeth, and travelled to Grahamstown, Port Alfred, King Williams Town and East London. Notes on the many plants he encountered, especially bulbous plants, cycads and proteas were published in *The Gardener's Chronicle* later that year (Gunn & Codd 1981).

In 1888, a specimen of a yellow-flowered form of *C. miniata* of undocumented origin was exhibited for the first time in Europe at a flower show in Ghent, and informally reported in a German gardening magazine as '*Clivia sulphurea* Laing & Sons' (Anon. 1888). During the same year, another yellow-flowered form was for the first time collected from a known locality in KwaZulu-Natal (Phillips 1931). The resident commissioner of Natal, Sir Melmoth Osborne, came across the plant amongst the firewood collected by a staff member in the Eshowe Forest, and two offsets of the original plant were obtained by Sir Osborne's assistant, Sir Charles Saunders, who in turn sent one offset to his artist mother, Katharine Saunders, some distance to the south in Tongaat (Thurston 1998). Plants were propagated from seed from these parents, and after many years produced yellow flowers, indicating a pure strain. The flowering specimen that Katharine Saunders painted had spent several days in transit via ox-wagon, and was therefore past its best when it reached her, but she completed a watercolour painting of it which was reproduced in the book *Flower Paintings of Katharine Saunders* (Bayer 1979). A quoted passage in this book, titled '*Clivia miniata* var. *flava*' accompanying Plate 20 reads as follows: 'Yellow Imantophyllum from Eshowe, flower withering after being two days in post bag. Most lovely, delicate, peculiar shade of yellow, not orange, but like straw colour mixed with pink, quite inimitable by me. October 8th, 1893. This drawing has been sent to Kew with the bulb by Maud'. Additional text states that the plant sent to Kew was successfully established and flowered there under glass in one of the greenhouses. It was on this material that the variety *flava* was based when E.P. Phillips published *Clivia miniata* var. *flava* E.P.Phillips in volume 11 of *The Flowering Plants of South Africa* (Phillips 1931), apparently unaware that the plant had already been described by Watson.

Watercolour of *Clivia miniata* var. *citrina* of cultivated origin, by Elbe Joubert

Watercolour of *Clivia caulescens* by Gilllian Condy, painted for the *Clivia* postage stamp series issued in 2006

Reproduced by courtesy of the South African Post Office

Curiously, Watson makes no mention of this collection from Eshowe, which must have been at Kew when he published his note and description of var. *citrina* (Watson 1899a, b).

In a report on *Clivia miniata* var. *citrina* in *The Garden*, William Watson (1899b) reported: 'Before the introduction of this new variety from Zululand we had only the Natal type, which has reddish orange flowers tipped with yellow on the lower half of the segments. This has been in cultivation nearly fifty years. According to Mr [J.G.] Baker [of Kew], it was introduced [into Britain] by Messrs. Backhouse and Sons in 1854, when a figure of it was published in *Curtis's Botanical Magazine* (t. 4783). Since that period many seedlings, more or less differing in the shade of red and form of the flowers, have been raised in English and continental gardens, and there are now numerous named sorts. A comparison of some of them with the type as represented in the figure above cited will show how little variation has so far been obtained. Much of the so-called variety of size and colour is due to cultural conditions. I have seen twenty named varieties grown together for two or three years, and at the end of that time they were all alike – *C. miniata* simply. In *The Garden*, vol. xxi., p. 358, along with a plate of 'Marie Reimers', the best seedling form of that period, will be found an excellent article by Mr James O'Brien on these plants (there called Imantophyllums), in which it is stated that about fifteen years after the introduction of the type by Messrs. Backhouse a variety called superba or maxima was imported, presumably from South Africa, by the late Mr Wilson Saunders, and that progress in new varieties may be said to date from the advent of that new variety, which differed from the type in having the flower segments almost white at the base. It is probable that the distribution of *C. miniata* extends considerably beyond the limits of Natal, and that colour variations of it will continue to be found as the countries adjacent to Natal become more explored'. Referring to the breeding potential of *C. miniata* var. *citrina* in the same article, Watson continued as follows: 'If the introduction of such a slight variation of the type as superba is resulted in the breeding of such varieties as 'Marie Reimers' and the newer 'Admiration', 'Favourite', 'Optima', etc., much more may be expected from crosses between the variety *citrina* and some of these. Apart, however, from its value as a breeder, the plant itself is a beautiful addition to greenhouse plants'.

More than 40 years later, a fourth *Clivia* species, the intriguing *C. caulescens*, another pendulous-flowered plant which develops a curious aerial stem with advanced age, from Swaziland and South Africa's Mpumalanga and Limpopo provinces, was described by Dr R.A. Dyer in *The Flowering Plants of South Africa* (Dyer 1943). A variable, lithophytic species, it flowers mainly in spring and early summer, although sporadic blooms may occur at any time of year (see page 72).

Notwithstanding just over four centuries of continuous botanical documentation of the geophyte flora at the Cape, beginning with bulbs of *Haemanthus coccineus* collected at the Cape in 1603 that flowered at Middelburg in The Netherlands in 1604, year after year the region continues to surprise us with the appearance of yet more species new to science. The discovery of *Clivia mirabilis*

Watercolour of *Clivia mirabilis* by Gillian Condy, painted for the *Clivia* postage stamp series issued in 2006
Reproduced by courtesy of the South African Post Office

by Johannes Afrika and Wessel Pretorius at Oorlogskloof Nature Reserve near Nieuwoudtville in the Northern Cape in 2001, a fifth species for the genus, caused a sensation. Not only was it morphologically distinct in numerous aspects, but its distribution was in a semi-arid region of the winter rainfall zone, in habitat quite unlike that of any previously described *Clivia*. It was published by Dr J.P. Rourke in *Bothalia* in 2002. It is interesting to note that *C. mirabilis* became only the second member of the genus after *C. caulescens* to be named in its country of origin.

In 1960, specimens of an unidentified, tubular-flowered *Clivia* were collected in swampy habitat north of Port St Johns in the north-eastern part of the Eastern Cape by W.L. Chiazzari, and deposited in the National Herbarium in Pretoria. Subsequent collections of the same plant were made in similar swampy habitat in this part of the Eastern Cape, and in southern KwaZulu-Natal. They were at first regarded as robust forms of *Clivia gardenii*, but subsequent investigations led a number of researchers to describe the plant as a new species, *C. robusta*, in the *Botanical Journal of the Linnean Society* (Murray et al. 2004). In addition to *C. miniata* var. *citrina* described in 1899, yellow forms of two more *Clivia* species have recently been described at varietal level, *C. gardenii* var. *citrina* from northern and central KwaZulu-Natal, published in *Bothalia* (Swanevelder et al. 2005) (see page 96) and *C. robusta* var. *citrina* from southern KwaZulu-Natal, published in Bothalia (Swanevelder et al. 2006) (see page 91).

Below: *Clivia* x *nimbicola*

Watercolour of *Clivia robusta* by Claire Linder Smith

Reproduced by courtesy of The Editor, *Flowering Plants of Africa*, Vol. 53 Plate 2094 (1994).

In addition, a new natural hybrid for the genus, *Clivia* x *nimbicola* (between *C. miniata* and *C. caulescens*) was described from the border between South Africa and Swaziland, in *Bothalia* (Swanevelder *et al*. 2006). In 1969, a number of plants thought to be *C. miniata* were collected outside their usual spring flowering season on The Bearded Man Mountain, a high peak on the border between eastern Mpumalanga and northern Swaziland, where an unusual blush-pink form of *C. caulescens* also occurs. They were planted in the Lowveld National Botanical Garden at Nelspruit and when they flowered, some of them produced unusual flared, pinkish tubular flowers, whereupon it became apparent that plants of a natural hybrid between *C. miniata* and *C. caulescens* had inadvertently been collected together with some of *C. miniata*. Several subsequent visits were made to The Bearded Man but no flowering hybrid plants were found until a surprise visit in midwinter 2003, when several of the hybrids were seen in full bloom, having flowers of a pale blush apricot, the first report of a natural hybrid for the genus (Rourke 2003a). The hybrid was recently named *Clivia* x *nimbicola,* the first formally described natural hybrid for the genus (Swanevelder *et al*. 2006) (see page 103).

With clivias having been cultivated for well over a century and a half, many artificial, interspecific hybrids and selections have been made, particularly in the United Kingdom, Belgium, Germany, Japan, The Netherlands and the USA. The most commonly encountered interspecific *Clivia* hybrid is *Clivia* x *cyrtanthiflora,* now more correctly known as *Clivia* Cyrtanthiflora Group (Koopowitz 2000, 2002), a hybrid between *C. nobilis* and *C. miniata.* It was first raised by Charles Raes in Ghent, Belgium in the late 1850s, and later figured by the highly successful plantsman and businessman Louis van Houtte (1810–1876) in *Flore des Serres et Jardins de l'Europe* in 1869, as *Imatophyllum* Cyrtanthiflorum, yet more than a century was to pass before it was

Clivia Cyrtanthiflora Group

finally transferred to *Clivia* by Hamilton P. Traub in *Plant Life* (Traub 1976).

Early pioneers of *Clivia* cultivation and breeding in South Africa were undoubtedly the inimitable Gladys Blackbeard and the intrepid Gordon McNeil, both of whom belonged to that rare breed of person (at that time), where individuality of spirit, and obsession with clivias and the environment, meant everything. Beginning in the late 1920s, Miss Gladys Blackbeard reared a fabulous collection of *Clivia* hybrids over a period of more than thirty years at Scott's Farm, Grahamstown in the Eastern Cape (Blackbeard 1939, Forrsman 1948).

Over a fifty-year period, Gordon McNeil amassed a vast collection of *Clivia* species and hybrids, as well as many other bulbous plants, at Cyprus Farm near Ofcolaco, in Limpopo, which he tended right up until his death in 1986. Gordon's *Clivia* breeding began in 1962 when he bought Gladys Blackbeard's collection which, according to Gordon's sister-in-law, Mrs Adelaide McNeil, 'required a whole railway truck to transport all the plants to the nearest railway station, and then to Cyprus Farm, where in ideal conditions they continue to thrive' (McNeil 1998). Gordon conducted countless hybridization experiments with his bulbs, including many intergeneric crosses; he was particularly proud of his putative intergeneric hybrid between *Clivia miniata* and an unidentified *Hippeastrum* species, which he named 'Green Girl', of which the author was fortunate enough to receive a plant shortly before Gordon's passing (McNeil 1985). Since his death, his clivias continue to be tended by his wife, Marguerite Rose McNeil, at Cyprus Farm in the remote and mountainous Legalametse Nature Reserve (Duncan 1999).

Below: Yoshikazu Nakamura in one of the shade houses at his Clivia Breeding Plantation, Chiba Prefecture, Japan, 1999

Towards the latter part of the 20th century, the focus on *Clivia* breeding shifted to the Far East, where a most impressive range of intraspecific hybrids (hybrids between different forms of *C. miniata*) as well as interspecific hybrids (hybrids between different *Clivia* species) were raised. *Clivia miniata* is a very popular pot plant in China and North Korea, and to a lessser extent, Japan. During a visit to Japan in 1991, the author was astonished to find a 120-page book by Ryo Ogasawara, illustrated in colour, in a local Kyushu supermarket, covering every imaginable aspect of its cultivation and propagation (Ogasawara 1997) and a new edition has just been published by his son, Sei Ogasawara (Ogasawara 2008). Masters of the art of plant selection, and seemingly obsessed with all plants exhibiting variegated foliage, the Japanese have produced a remarkable array of dwarf and variegated forms of *C. miniata*, in addition to numerous interspecific hybrids. Most famous among present-day *Clivia* breeders in that country is the affable and generous Mr Yoshikazu Nakamura, who holds the world's most diverse collection of *Clivia* germ plasm at his 'Clivia Breeding Plantation' south of Tokyo. Mr Nakamura became acquainted with Mr Isamu Miyake, a breeder of *Alstroemeria* and owner of a very successful nursery in the same area of Chiba Prefecture, who specialized in exotic plants. Nakamura was introduced to Dr Hirao by Miyake and eventually, after Hirao's untimely death, Nakamura inherited Hirao's *Clivia* collection, giving Nakamura an excellent head start in assembling his own collection, which formed the basis of his Clivia Breeding Plantation. Known in Japan as *kunshi-ran*, the popularity of *Clivia miniata* there is curiously still in its infancy.

Equally popular, if not more so, is the cultivation and breeding of *C. miniata* in the People's Republic of China, where dwarf, orange-flowered cultivars and variegated plants are widely grown as indoor pot subjects. First introduced there from Europe in the latter half of the 19th century, and again in the early 1930s by the Japanese, it became a treasured container plant inside the palaces of the last Imperial Ching dynasty due to its symbolic longevity, with showy leaves further enhanced by flowers in season. The industrial city of Changchun, situated in China's north-eastern Jilin Province, has become the centre of *Clivia miniata* development in that country. This came about after it was designated the capital city of Manchuria following the Japanese invasion of this territory, where the last Chinese Emperor Pu Yi was stationed as a puppet ruler during the occupation (Behr 1987). It can confidently be assumed that *Clivia miniata* reached Changchun from Japan, and its popularity in Changchun reached such levels that its flower became the city's emblem in October 1984. At just over 200 years old, Changchun is relatively young when compared with other Chinese cities, but already it has a Clivia Industrial Office, a Clivia Society, and a Clivia Academic Committee.

Now grown in many parts of China, special *Clivia* selections are regarded as status symbols there, and considered a sound investment. During a visit to that country in the late 1990s, the author was greatly surprised to see countless pots of dwarf-flowering *Clivia miniata* hybrids surrounding the embalmed body of Mao Tse Tung inside the Chairman Mao Memorial Hall on Tiananmem Square, Beijing, which greatly relieved the otherwise sombre, austere surroundings. The cultivation of *C. miniata* in the Far East is focused primarily on the beauty of the dark green, shiny leaves, as well as those with variegated foliage, which provide enjoyment throughout the year, not only when in flower. Chinese growers pay extra special attention to the leaves, showing virtually no interest in the flowers. In the Chinese sense, the leaves of a good specimen must be stacked as symmetrically as possible, they must be short and broad with rounded tips, and shiny and thick with dark green, protruding veins, preferably on a yellowish background. In contrast to Japanese breeding, which places great emphasis on downwardly curved leaves, the Chinese aspire towards upwardly oriented leaves. In some newly developed Chinese plants, the breadth of the leaves is almost equal to their length, and they have given their special plants their own cultivar names like 'Yellow Engineer', 'Buddhist', 'Victoria' and 'Painter'.

Illustrating the worldwide popularity of clivias, their images have been featured on the postage stamps of at least ten countries, including Burundi (*C. miniata*), People's Republic of China (*C. nobilis* and three forms of *C. miniata*), Lesotho (*C. nobilis*), North Korea (*C. miniata*), Republic of China [Taiwan (*C. miniata*)], Rumania (*C. miniata*), Saharawi [Western Sahara (*C. miniata*)], South Africa [*C. caulescens, C. gardenii, C. miniata, C. mirabilis, C. nobilis* and *C. robusta*, including the former Independent Homelands of Ciskei (*C. nobilis*) and Venda (*C. caulescens*)], The Maldives (*C. miniata*) and Togo (*C. nobilis*) (Nel 2006a, b). South Africa's set of stamps, superbly illustrated by Gillian Condy, the resident botanical artist at the South African National Biodiversity Institute in Pretoria, was issued by the South African Post Office in September 2006, coinciding with the 4th International *Clivia* Conference held in Pretoria (Dixon 2006b).

A tremendous resurgence in the cultivation, propagation, breeding and study of *Clivia* has taken place over the past fifteen years, which has progressed to an international obsession with the genus. In South Africa it resulted in the formation of the *Clivia* Club in 1994, which in 2001 became the *Clivia* Society that currently includes nine regional clubs and five interest groups across the country, and enjoys an impressive local and international membership of at least 1500. It also has international representatives in Australia, New Zealand, The Netherlands, United Kingdom and United States of America (see page 186).

Specialist *Clivia* nurseries have arisen in a number of countries and clivias are now marketed around the world, having

become sought after items of merchandise, notably in China where outrageous sums are often realized for especially well bred specimens.

Since the publication of *Grow clivias* in 1999, our knowledge of *Clivia* has grown exponentially. Two new species, two new varieties and one new natural hybrid have been described. Numerous scientific studies including ploidy research, breeding, phylogeny, population structure and pigment analysis have been undertaken, and numerous successful international conferences have taken place. The number of intraspecific and interspecific hybrids have increased markedly, and a new, artificial interspecific hybrid between *C. miniata* and *C. mirabilis* flowered for the first time in 2006 (see page 110). Pastel-coloured hybrids of *C. miniata* are slowly becoming more readily available to the public, and many new cultivars have been officially registered (see page 103). In addition, many highly successful *Clivia* shows continue to take place across South Africa and in several countries abroad; the Chinese already have an established quality standard and judging system in place for their shows, while the *Clivia* Society's *Handbook on judging, showing and registration* is gradually being fine-tuned into an indispensable document for use in South Africa.

The Clivia First Day Cover issued by Philatelic Services in 2006

Reproduced by courtesy of the South African Post Office

Below: *Clivia miniata* 'Kirstenbosch Supreme' in the Kay Bergh Bulb House of the Botanical Society Conservatory at Kirstenbosch (see page 116)

Opposite: *Clivia robusta* from the Eastern Cape (see page 88)

GENERAL INFORMATION

Introduction

Appreciated around the world for their long-lasting showy flower heads and ease of culture, the endemic southern African genus *Clivia* Lindl. comprises six species belonging to the family Amaryllidaceae. The species are *C. caulescens* R.A.Dyer, *C. gardenii* W.J.Hooker, *C. miniata* (Lindl.) Regel, *C. mirabilis* Rourke, *C. nobilis* Lindl. and *C. robusta* B.G.Murray *et al*. Within three of these species, naturally-occurring yellow-flowered variants are recognized, *C. gardenii* var. *citrina* Z.H.Swanevelder *et al.*, *C. miniata* var. *citrina* Watson and *C. robusta* var. *citrina* Z.H.Swanevelder *et al*. A recently published natural hybrid between *C. miniata* and *C. caulescens*, *C.* x *nimbicola* Z.H.Swanevelder *et al*. is also recognized. While there can be no mistaking the large, trumpet-shaped flowers of *C. miniata*, it is the five tubular-flowered species which, at a glance, look very similar. Superficially, they resemble certain members of the genus *Cyrtanthus* L.f. (such as *C. herrei* and *C. obliquus*) in their similar flower shape and colouring, but are distinguished mainly in having rhizomes instead of true bulbs, much shorter perianth tubes, and fruits producing large globose or

Opposite, left to right, top to bottom: *Clivia caulescens, Clivia gardenii* var. *gardenii, Clivia miniata* var. *miniata, Clivia mirabilis, Clivia nobilis* and *Clivia robusta* var. *robusta*

Below, left to right, top to bottom: *Clivia gardenii* var. *citrina, Clivia miniata* var. *citrina, Clivia robusta* var. *citrina* and *Clivia* x *nimbicola*

subglobose, water-rich seeds inside a fleshy berry, in contrast to papery dormant seeds produced inside a capsule as in *Cyrtanthus*. The five tubular-flowered clivias can usually be distinguished from one another by a combination of the following features: leaf apex shape, leaf margin texture, presence or absence of an elongated aerial stem, pedicel colour at flowering, position of the stamens and stigma in relation to the perianth apex and size of the mature seed. However, it should be borne in mind that due to natural variation within each of these species, the above-mentioned features are not always constant. The natural distribution and flowering period can also be used as an additional aid to identification (see page 56 for a key to the genus and pages 58-98 for information on individual species).

Phylogeny is the pattern of evolutionary history and common ancestry among species, and cladistics is the method used to reconstruct genealogies of these species, and to construct natural classifications based on ancestry. The most frequently employed method of achieving this is the use of DNA sequences and, to a lesser extent, morphological characters. Another method used in hypothesizing relationships lies in genome size measurement; when all species in a genus have the same chromosome number, as in *Clivia* ($2n = 2x = 22$). This method has shown that differences in nuclear DNA content, when evaluated in conjunction with morphological data, can be very effective in delimiting groups of species in a number of genera, as has been shown, for example, in *Agapanthus* (Zonneveld & Duncan 2003) and *Nerine* (Zonneveld & Duncan 2006). Results of a recent analysis of chloroplast DNA sequences of five *Clivia* species (excluding *C. robusta*) showed a close relationship between the summer rainfall species, with the enigmatic *C. mirabilis* from the winter rainfall area being the sister to all the other species. Moreover, the trumpet-shaped flowers of *C. miniata* were shown to be derived from the basic tubular floral form (Conrad & Reeves 2002). These findings correlate well with the results of a study of genome size in *Clivia* species, which showed that nuclear DNA amount differed in all six species, varying from 31.2 to 39.2 picograms (pg), with *C. mirabilis* having the lowest DNA content (Zonneveld 2002, 2005). The fact that the nuclear DNA values do not differ greatly between the species, and that all the species can be cross-fertilized and produce fertile offspring, suggests that they are all closely related to each other. *Cryptostephanus*, the genus most closely related to *Clivia*, is an entirely African group comprising three species, *C. haemanthoides* from Kenya and Tanzania, *C. vansonii* from the mountains of eastern Zimbabwe and *C. densiflorus* from Angola and northern Namibia. The only one of these cultivated to any extent is the evergreen, white- or rarely pink-flowered *C. vansonii* that occurs in similar habitat to the summer rainfall clivias, found in decaying leaf litter in dappled shade among rocks on the forest floor. *C. densiflorus* and *C. haemanthoides* are deciduous, summer-growing species with greyish-green leaves and dense heads of reddish-purple flowers, occurring in dry, or moist, shaded or semi-shaded woodland habitat, and are rare in cultivation (Duncan 2000b).

Above: Ripe berries of *Cryptostephanus vansonii*
Right and below: *Cryptostephanus vansonii* in flower

Growth and flowering cycles

Clivias are long-lived, slow-growing plants. All six species are evergreen with perennial thick fleshy roots that spread out horizontally and are very well equipped for water storage, allowing the plant a remarkable level of drought resistance. Certain forms of *C. robusta* found growing in swampy conditions are anomalous in producing buttress-like roots that develop from the base of the pseudostem as it rises above the water line, and in the lithophytic *C. caulescens*, roots often develop along the aerial portion of the rhizome. The roots in all *Clivia* species are covered with a corky, water- and nutrient-retaining velamen layer that is especially thick in *C. mirabilis*, enabling it to survive the harsh summers in its semi-arid, winter rainfall habitat. The remaining five species occur naturally in areas that normally receive summer rainfall and minimal, if any, winter rainfall. Although evergreen, the summer rainfall clivias naturally undergo a dormant period during the dry winter months when few, if any new leaves are produced.

C. nobilis is the most slow-growing species, even when in active growth, taking many years to flower in cultivation when grown from seed. With the onset of spring and summer rain, active growth begins in the summer rainfall species, with new leaves produced at the apex of the growing shoot, while the oldest, outer leaves turn yellow, then brown, and finally detach from the plant. In *C. mirabilis*, the only winter rainfall species, the active growth cycle is reversed, new leaves beginning to form in autumn with

the onset of cooler weather, continuing throughout winter, and ceasing with the onset of warmer weather in spring. As a group, clivias have a long flowering period in their natural habitat, beginning in mid autumn and continuing to early summer. Viewed as separate species, *C. gardenii* and *C. robusta* flower from mid autumn to midwinter, *C. nobilis* flowers from midwinter to early summer, *C. miniata* flowers from late winter to early summer, *C. caulescens* flowers mainly in spring and early summer, and *C. mirabilis* flowers from late spring to early summer. With the exception of *C. gardenii*, *C. mirabilis* and *C. robusta*, sporadic blooms may appear in the remaining three species at any time of year under cultivation, with *C. caulescens* and certain forms of *C. nobilis* frequently producing a second flush of blooms in autumn, and certain forms of *C. miniata* var. *citrina* sometimes producing sporadic blooms in early winter.

Stems and foliage

The organ from which the thick adventitious, branching fleshy roots of clivias develop is usually a short, subterranean, vertically oriented rhizome, similar to that of *Agapanthus*. New leaves are produced annually from the centre of the growing shoot, while several of the older, outer leaves die off each year (Du Plessis & Duncan 1989). In mature specimens of *C. caulescens*, the rhizome extends above ground level to form a distinct aerial stem up to 1 m long or more. In the wild, all the *Clivia* species are offset-forming with the exception of *C. mirabilis*. However, under cultivation the latter species does sometimes produce offsets. In certain forms of *C. gardenii, C. miniata* and *C. miniata* var. *citrina*, one or more subterranean stolons may develop from the top of the rhizome, just below ground level, each giving rise to a new plant.

Opposite, left: Root development along the aerial portion of a rhizome in *Clivia caulescens*

Opposite, right: New leaf development in *Clivia mirabilis*

Left: Offset formation in *Clivia mirabilis* in cultivation

Clivias all have strap-shaped leaves arranged in a pseudostem in two opposite rows, and vary from leathery and sub-erect to erect in C. mirabilis and C. nobilis, to soft-textured and spreading to arching in the other four species. Unique within the genus are the leaf margins of C. nobilis in having a cutting edge due to minute serrations. Leaf colour, length and apex shape varies considerably among the different species in their natural habitat, sometimes even within different forms of the same species. For example, there is variation from the dark green, arching, weakly canaliculate, smooth, plain green leaves of C. gardenii, to the glaucous, sub-erect to erect, leathery, often deeply canaliculate leaves of C. mirabilis. The leaves of forms of C. nobilis occurring in the deep shade of inland forests are dark green and can grow up to 1 m long, whereas those occurring on primary and secondary dunes in light shade or full sun tend to be glaucous, and considerably shorter and broader. Certain forms of C. caulescens have broad, pale green arching leaves with an obtuse-acute apex, while others have relatively narrow, dark green sub-erect leaves with an acute apex. In most forms of C. nobilis the leaf apex is rounded, or notched to a greater or lesser degree, and there is sometimes considerable variation in the leaf tip shape in leaves of the same plant of this species. The broadest and longest leaves are found in certain forms of C. robusta that grow up to 1.2 m long and up to 90 mm wide, and its leaf apices vary from obtuse apiculate to obtuse acute. Greatest variation in leaf form is evident in C. miniata, ranging from narrow with an acute apex to relatively broad with a sub-acute apex, but when one considers the wide natural distribution of this species, extending from the former Transkei in the Eastern Cape through KwaZulu-Natal to Swaziland and Mpumalanga, this is hardly surprising. The new leaves of most forms of C. mirabilis are distinctive in having a broad white striation running down

the entire central portion, fading as the leaf matures. Similar leaf striation is also sometimes seen in forms of the closely related *C. nobilis*.

In addition to natural variation within the leaves of clivias in the wild, a vast array of striking foliage types has arisen over the past century and a half in cultivation, as a result of intensive selection and hybridization. In this respect one considers the large, broad-leafed intraspecific hybrids of *C. miniata* developed in Belgium in the late nineteenth and early twentieth centuries, that were forerunners to the dwarf, broad-leafed 'Daruma' hybrids raised in Japan, as well as countless broad-leafed hybrids of *C. miniata* that have been raised in China. Spontaneous genetic mutation sometimes occurs in the leaves of clivias, in the wild as well as under cultivation, resulting in the appearance of irregular, longitudinal stripes or blotches, called variegation. The colour of these stripes and blotches ranges from pure white through many shades of cream to yellow and greenish-yellow, or a combination of these. Variegation in *Clivia* leaves, which is most frequently encountered in *C. miniata*, is regarded as highly decorative and sought-after by certain members of the *Clivia*-growing community, and in China and Japan this phenomenon is regarded as particularly desirable. In fact, Japan's most well-known contemporary *Clivia* breeder, Mr Yoshikazu Nakamura, concentrates on the production of clivias with variegated leaves in order to satisfy local demand, and even employs artificial irradiation treatments on his plants in order to induce a wide range of variegation patterns.

Opposite, left: Variation in *Clivia nobilis* leaf apex morphology

Opposite, right: Median striation in leaves of *Clivia mirabilis*

Below: 'Shima-fu' variegated *Clivia miniata*

Inflorescence and flowers

The inflorescence in *Clivia,* like that of *Agapanthus,* is strictly speaking a pseudo-umbel (Müller-Doblies 1980, Rourke 2003b). It is borne atop an elongated, solid, leafless stem, known as the scape, and this is laterally compressed, having two clearly defined edges. It develops from one of the leaf axils, and always appears laterally, i.e. not from the centre of the plant. The flowers are attached to the scape apex by pedicels or flower stalks, and in *C. miniata* they are thick and radiating, while those of all the pendulous-flowered species vary from drooping to erect and are relatively narrow. The number of flowers per pseudo-umbel varies greatly among the species, and among different forms of the same species. The minimum number of flowers per head is ten, seen in *C. gardenii* and *C. robusta*, increasing to as many as 50 in robust forms of *C. caulescens* and *C. nobilis*. Pedicel colour in *C. gardenii,* *C. miniata, C. nobilis* and *C. robusta* is always green, whereas in *C. mirabilis* it changes from green to bright orange-red or purple during anthesis, and back to green during the fruiting stage. Contrary to recent literature (Rourke 2002a), *C. mirabilis* is not the only species having pedicels the same colour as the perianth, as certain forms of *C. caulescens* in eastern Mpumalanga also exhibit this trait (see next page).

The immature inflorescence is enclosed by four or more membranous spathe bracts that become papery with age, and some of the outermost flowers have thread-like secondary bracts called bracteoles. In rare instances, one or more secondary pseudo-umbels may develop from the top of the scape in *Clivia* and bear one to several flowers, a phenomenon occasionally seen in *C. caulescens* and *C. miniata*, and also recorded in *C. mirabilis* (Rourke 2003b). This phenomenon, seen quite commonly in *Agapanthus praecox*,

occurs both in the wild and under cultivation. It is not yet known whether the development of secondary pseudo-umbels in *Agapanthus* and *Clivia* is an entirely random occurrence as a result of environmental conditions, or whether it is an inherited trait.

Opposite: Orange-red and purple pedicel colouration in *Clivia mirabilis*

Right: Orange-red pedicel colouration in *Clivia caulescens*

Below: Development of secondary pseudo-umbel in *Clivia caulescens* in habitat

Below, right: Spathe bracts of *Clivia mirabilis*

Clivia floral parts

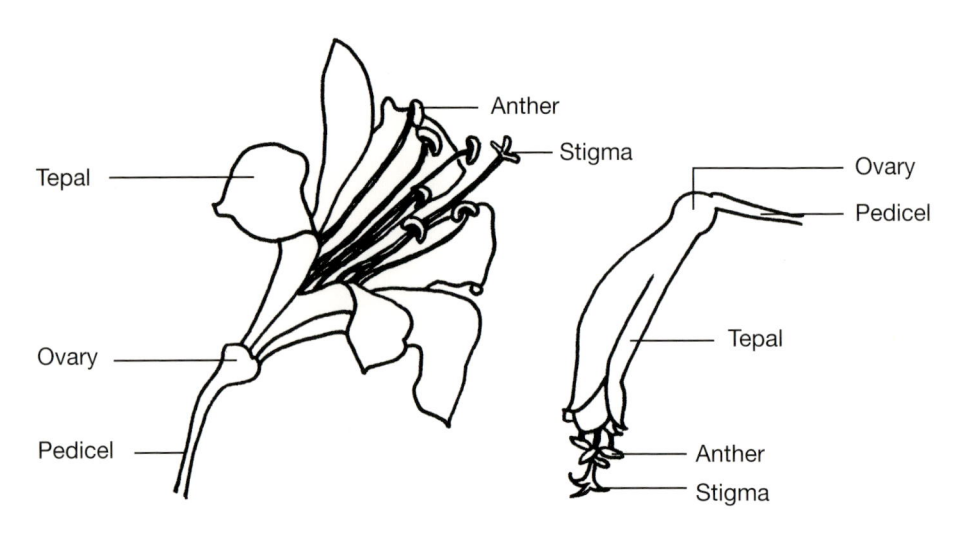

Tepal — Anther — Stigma — Ovary — Pedicel — Ovary — Pedicel — Tepal — Anther — Stigma

On the basis of flower form, the genus *Clivia* can be divided into two groups of species, one having spreading to erect, trumpet-shaped flowers *(C. miniata)* above left, the other having narrow, tubular, usually pendulous flowers *(C. caulescens, C. gardenii* (right), *C. mirabilis, C. nobilis* and *C. robusta)* above right.

Infloresence
Scape
Pseudostem
Leaf sheath
Offset
Rhizome
Stolon

Clivia miniata exhibits a remarkable flower colour range in the wild, from pale yellow (*C. miniata* var. *citrina*), through many shades of orange to dark red, and there are also several naturally occurring forms with pastel coloured flowers, ranging from apricot to peach pink. Flower colour in most populations varies in shades of pale to deep orange, and in some populations deep reds and pastel colours occur amongst the oranges. Even within certain wild populations of this species occupying a small surface area, a high level of variation in flower colour exists; clearly such populations are evolving at a much faster rate than others. In the absence of pollination, *Clivia* flowers are long lasting, remaining attractive for up to approximately 18 days. It should be borne in mind that flower colour in a number of forms of *C. miniata*, *C. nobilis* and *C. mirabilis* changes with age, or due to environmental conditions. In most forms of *C. miniata* var. *citrina*, flowers open pale creamy-yellow or greenish-yellow, darken somewhat when in full flower, then fade to a washed out, yellowish-white before dropping off. Flower colour of the species with tubular flowers varies from pale to dark orange or orange-red, occasionally pale yellow (*C. caulescens*, *C. gardenii* var. *citrina*, *C. nobilis* and *C. robusta* var. *citrina*) or pale pink (*C. caulescens*, *C. gardenii*, *C. mirabilis* and *C. nobilis*) and the flowers are almost always tipped with pale to dark green, or rarely with yellow.

Below: *Clivia miniata* var. *citrina* 'Natal Yellow' (left) and *C. miniata* hybrid (right)

Above: The trumpet-shaped flowers of *Clivia miniata*

Opposite, left: Azalea-scented flowers of *C. miniata* var. *citrina* 'Kirstenbosch Yellow'

Opposite, right: Tubular flowers of *C. mirabilis*

The perianth of the *Clivia* flower is actinomorphic (regular) in *C. miniata* and *C. mirabilis*, weakly zygomorphic (irregular) in *C. caulescens, C. nobilis* and *C. robusta,* and strongly zygomorphic in *C. gardenii*. It consists of a perianth tube that is funnel-shaped in *C. miniata* but tubular in the five tubular-flowered species, and three outer and three inner tepals. *Clivia mirabilis* (10–15 mm) and *C. robusta* (10–14 mm) have the longest perianth tubes, followed by *C. gardenii* (7–10 mm) and *C. miniata* (7–10 mm), with *C. caulescens* (6–8 mm) and *C. nobilis* (5–7 mm) having the shortest tubes. The outer tepals are distinctly narrower and slightly shorter than the inner tepals in all the *Clivia* species, the outer tepals varying from linear to broadly lanceolate, and the inner tepals from spathulate to oblanceolate. Within the perianth there are six stamens, each consisting of a linear, straight or slightly curved white filament with an oblong, dorsifixed and versatile anther at its tip. The ovary is situated at the base of the flower, below the tepals and stamens, and gives rise to a linear white style with a tricuspidate stigma at its tip, that usually extends beyond the anthers. The ovary has three locules or compartments, each containing up to nine ovules per locule.

Certain forms of *C. miniata* have a fairly strong scent reminiscent of azaleas, and is strongest in *C. miniata* var. *citrina* 'Kirstenbosch Yellow' (Duncan 1999). The stamens and style of the *C. miniata* flower are usually shorter than the tepals, while those of *C. caulescens* are as long as the tepals, or protrude slightly beyond them. In *C. nobilis* the style and stamens may be included or protrude up to 6 mm beyond the tepals. *C. gardenii* usually has conspicuously exserted styles and stamens, but in *C. robusta* they are usually included or just emerging beyond the perianth. *C. mirabilis* has included stamens and an included or slightly exserted style.

Fruits and seeds

Following successful pollination and fertilization, the ovules in the ovary develop into seeds that form inside the fruit that develops at the tips of the pedicels. The individual fruit containing the seeds is a berry, and consists of a yellowish inner pulpy layer surrounded by a brightly coloured pericarp (outer skin). Berry shape varies from globose to subglobose in most species but those of *C. mirabilis* are unusual in being irregularly oblong or ovoid, with their apices often narrowed to a narrow, beak-like structure. The number of seeds within each berry varies tremendously, from one to as many as 25 seeds (as in certain forms of *C. miniata*). The colour of the embryo is cream, surrounded by white endosperm. It should be borne in mind that frequently one or more ovules may abort at an early stage of development, resulting in fewer seeds. In general, it is the berries of *C. miniata* which contain the most seeds. *C. gardenii* and *C. robusta* have the largest seeds (approx. 18 mm diam.), followed by those of *C. miniata* (approx. 15 mm diam.), *C. caulescens* (approx. 12 mm diam.), *C. mirabilis* (approx. 10 mm diam.), while *C. nobilis* has the smallest seeds (approx. 9 mm diam.). A ripening *Clivia* berry is normally green, while those of many variegated clivias often also produce variegated berries. The *Clivia* fruit takes between five and 12 months to ripen, but will often remain on the plant well past maturity, if left to its own devices. With the exception of the winter-growing *C. mirabilis,* the berries of which ripen rapidly from late summer to early autumn in order to coincide with autumn rains, those of all the other species ripen from mid to late winter to coincide with spring rains. Eventually the outer fleshy layer disintegrates and releases the hard, pale brown or cream seeds, which often begin to germinate while inside the fruit wall. The berries of *C. gardenii, C. miniata* and *C. robusta* take from nine to 12 months to mature, while those of *C. caulescens*

Opposite: Fruits of *C. mirabilis* (left) and variegated *C. miniata* (right)

Above: Fruits of ***C.** miniata* (left) and *C. gardenii* (right)

Below: Seeds of *C. miniata* (left) and fruits of *C. caulescens* (right)

Above: Fruits of *C. gardenii* (left), *C. nobilis* (right) and *C. miniata* var. *citrina* (below)

and *C. nobilis* usually mature faster, from six to eight months, and those of *C. mirabilis* mature in as little as five months.

At maturity, the colour of the berry varies from pale to dark red, orangy-red, yellow, or a mixture of red, yellow and green, depending on the particular species or colour form. All forms of *Clivia miniata* that have flowers in shades of orange or red will produce orangy-red or red berries, while most forms of this species with cream or yellow flowers will produce uniformly yellow berries. The berries of certain forms of yellow-flowered plants such as *C. miniata* var. *citrina* 'Natal Yellow' may be marked with dull reddish spots, or exhibit a mixture of red, yellow and green. Forms of *C. miniata* with peach-coloured flowers produce either peachy-yellow or red fruits. *C. caulsecens, C. gardenii, C. mirabilis* and *C. robusta* var. *robusta* produce berries in shades of red or orange-red. Most forms of *C. nobilis*

have red berries but in certain forms yellow berries are produced.

True-breeding yellow forms of *C. miniata* var. *citrina* produce yellow-flowered offspring when their flowers are cross-pollinated, but it is important to note that most forms of *C. miniata* var. *citrina* in cultivation today are not true-breeding; harvesting of seeds from yellow-flowered plants does not necessarily mean that the offspring from such seeds will all produce yellow flowers. It depends entirely on the genetic make-up of the yellow-flowered parent plant from which the seeds have been harvested, as well as that of the pollen parent with which the flowers have been pollinated, as to the number of yellow-flowered plants (if any) that will be produced. Forms of *C. miniata* var. *citrina* have been categorized into two groups based on breeding behaviour, those belonging to 'Group 1' are true-breeding and those belonging to 'Group 2' are not true-breeding (Morris 1999) (see page 64). The ripe berries of *C. gardenii* var. *citrina* are anomalous in not being yellow, but bright orange-red to deep red, while those of *C. robusta* var. *citrina* vary from pale to dark yellow. Although the water-rich seeds of *Clivia* are much slower to germinate than many other southern African amaryllids, they are non-dormant and have a thin testa (seedcoat), germinating as soon as favourable conditions are present. They frequently germinate inside the ripe fruit before it has dropped to the ground.

Left: The ripe berries of *C. gardenii* var. *citrina* are anomalous in not being yellow, but bright orange-red to deep red

Pollination

The two different flower forms encountered in *Clivia* (trumpet-shaped and tubular) are adaptations to attract different types of pollinators. The flowers attract their pollinators primarily by colour and shape, and offer rewards of nectar, and to a lesser extent, pollen (*C. miniata*). It should also be borne in mind that, like the genus *Amaryllis*, clivias are at least partially self-fertile (Duncan 2004a, b), usually resulting in the production of a few seeds per infructescence, regardless of visits by potential pollinators.

Indications of probable sunbird pollination in tubular-flowered *Clivia* species became apparent when lesser double-collared sunbirds (*Nectarinia chalybea*) were observed gathering nectar in the Kirstenbosch bulb nursery from the three tubular-flowered species recognized at that time, *C. caulescens*, *C. gardenii* and *C. nobilis* (Duncan 1999). Subsequently, pollination by the olive sunbird (*Nectarinia olivacea*) of *C. gardenii* in cultivation in KwaZulu-Natal was observed by Val Thurston (Koopowitz 2002), as well as observations in the wild of the black sunbird (*Nectarinia amethystina*) visiting the flowers of *C. caulescens,* and the malachite sunbird (*Nectarinia famosa*) visiting the flowers of *C. mirabilis* (Manning 2005). Copious nectar is produced in *C. nobilis*, yet curiously, no documented pollination records exist for this species or *C. robusta.* In the tubular-flowered species, the sunbird clings to the sturdy scape, inserts its curved beak into the flower and in doing so, lifts

Below: The two different flower forms encountered in *Clivia* (trumpet-and tubular-shaped, respectively) attract different types of pollinators

the pendent flower into a more or less spreading position, as occurs in certain long-tubed species of *Lachenalia* that are also sunbird-pollinated such as the bright red *L. bulbifera* and the brightly coloured flowers of *L. aloides* (Duncan 2005). The sunbird probes the flower for nectar which collects in a compartment at its base, below the area where the base of the filaments arise from the perianth tube. In doing so, pollen is deposited onto the base of the beak and the forehead, which comes into contact with the stigma when the bird visits another flower and pollination takes place. The similarity in flower shape and colour of the tubular-flowered *Clivia* species with other southern African geophytes having pendent orange, red or yellow flowers that are known or deduced to conform to the bird pollination syndrome, like *Lachenalia aloides* (bright yellow or orange, or a combination of different bright colours), and *Cyrtanthus herrei* (reddish-orange with green tips), is a remarkable instance of convergent evolution.

The erect to spreading, trumpet-shaped flowers of *C. miniata* are visited by swallowtail butterflies (*Papilio demodocus*) in the wild (Snijman 2002b). These large, colourful butterflies are attracted to bright orange or red flowers and seem to be its principal pollinators. In the wild their larvae feed on members of the Rutaceae (citrus family) and are also commonly seen laying their eggs on the leaves of a range of exotic citrus trees in gardens throughout South Africa, especially lemons (*Citrus limon*) and kumquats (*Fortunella japonica*).

Below: Other southern African geophytes having similar tubular-shaped flowers like *Lachenalia aloides* (left) and *Cyrtanthus herrei* (right), also conform to the bird pollination syndrome in a remarkable instance of convergent evolution

During the course of nectar-feeding, the wings of the butterfly become dusted with pollen that is then transferred to the stigmas of other flowers, effecting pollination. In addition to butterflies, *C. miniata* is also visited by generalist insect pollinators under cultivation, including honey bees (*Apis mellifera*) that feed on the pollen by day, night-flying moths, and occasionally by sunbirds. Honey bees have been seen entering and proceeding to the base of the flowers of *C. miniata* var. *citrina* (Duncan pers. obs.). The flowers of certain forms of *C. miniata* var. *miniata* as well as *C. miniata* var. *citrina* have a fairly strong scent reminiscent of azaleas, one of the most strongly-scented yellow forms being *C. miniata* var. *citrina* 'Kirstenbosch Yellow' (Duncan 1999). The scent intensifies at night, and a common noctuid moth (*Trichoplusia* species) has been observed flying into the flowers of orange-flowered, cultivated specimens with deep yellow throats that were flowering out of season. According to Prof. Clarke Scholtz, up until 2004 there were no known records of moths feeding on *Clivia* and suggested the flowers of *C. miniata* may have a special attraction for the moths, such as some pheromone-like chemical (Fisher 2004). Whether or not a pheromone-like chemical is associated with the azalea-like scent is as yet unknown.

Distribution and habitat

Clivia is endemic to the southern African countries of South Africa and Swaziland. Its distribution is concentrated in the summer-rainfall zone in the eastern and north-eastern parts of the region, extending from the southern part of the Eastern Cape in a north-easterly direction, through southern, central, eastern and northern KwaZulu-Natal to northern Swaziland, and along the north-eastern escarpment of Mpumalanga to the Soutpansberg Mountains in Limpopo (Duncan 1999, Vorster & Smith 1994).

C. mirabilis is a disjunct species with a restricted range in the winter rainfall zone near Nieuwoudtville in the Northern Cape, and near Vanrhynsdorp in the Western Cape (Rourke 2002a, Grebe 2005, 2006). Altitudinal range for the genus is from sea level (*C. miniata*, *C. nobilis*, *C. robusta*) to approximately 1770 m above sea level on the north-eastern escarpment (*C. caulescens*).

The habitat frequented by the five summer rainfall species is primarily humid coastal and inland evergreen afromontane forest, where they mostly grow on south- and south-east facing slopes in varying degrees of shade. Occasionally, they may be found in situations receiving full sun for short periods. Populations usually occur in isolated patches, often widely separated from each other, usually in association with large boulders.

The most southerly species, *C. nobilis* is restricted to the southern and eastern parts of the Eastern Cape. It occurs in large colonies on relatively dry primary and secondary coastal dunes as well as in moister coastal and inland afromontane forest. It is the second-most sun- and drought tolerant species after *C. mirabilis*. In coastal dunes it grows amongst low scrub cover in deep sand, sometimes very close to the high water mark, where the plants occur in light shade or are exposed to full sun for short periods. However, in coastal and inland forest it is found in dense shade, mainly in decaying leaf litter on the forest floor, or occasionally in loam or shale soils, or in leaf litter deposits on large boulders. It is frequently seen growing in association with *Dracaena aletriformis* (forest dragon tree), *Rumohra*

The most northerly occurring species, *C. caulescens* is found in northern and central Limpopo, northern and eastern Mpumalanga and northern Swaziland (see page 72)

adiantiformis (seven-weeks fern) and *Strelitzia nicolai* (wild banana).

Considerable variation in flower colour is encountered across its range. In the inland western parts it tends towards pastel shades of pink or reddish pink, while in forest conditions further east, colours range in shades of pale red or orange, and in exposed coastal dunes they tend toward more intense shades of reddish-orange (Cowley 2006b). As is often the case in cultivation, flowering of *C. nobilis* in the wild is somewhat erratic, with relatively few individuals producing inflorescences simultaneously, compared with the number of mature plants in a given population. In the northern part of its range, *C. nobilis* is sympatric with *C. miniata*, but due to their different pollinating agents, reports of natural hybrids are extremely rare and no formally documented records exist.

Having the widest distribution, *C. miniata* is not surprisingly the most variable species, and it occurs from the eastern part of the Eastern Cape north-eastwards along the coast of this province and slightly inland, to the coastal parts of KwaZulu-Natal, the Midlands and northern part of this province, northwards into western and northern Swaziland, to eastern Mpumalanga. It is found as isolated colonies in afromontane forest patches, on dolerite and sandstone (Winter 2000). Wherever it has not been decimated by 'muthi' collectors, it is found in large colonies in decaying leaf litter on the forest floor, along shaded watercourses, on ledges and in ravines, occasionally in leaf litter on rocks, or

The forested habitat of *C. nobilis* (left) and *C. miniata* (opposite)

Cameron McMaster

even in the forks of trees (Duncan 1999). Flower colour, shape and size is uniform within most populations and shades of orange predominate, while in others flowers are almost red or occasionally pale yellow, or vary in a range of pastel colours including pink and apricot. Fruit and seed size is considerably larger in the more westerly populations (Winter 2000). Forms of the var. *citrina* have been recorded throughout the range of var. *miniata*, while those with pastel colours are encountered most frequently in the Eastern Cape. Like certain forms of *C. nobilis*, *C. miniata* it is often seen growing in association with *Dracaena aletriformis*. In addition to *C. nobilis*, the distribution of *C. miniata* overlaps with that of *C. robusta, C. gardenii* and *C. caulescens* in parts of its range, and in certain instances, mixed populations occur.

Considered by some authorities to be a separate species, but by others merely as a robust form of *C. gardenii*, *C. robusta* occurs in light to deep shade of afromontane forest in the northeastern part of the Eastern Cape and the southern part of KwaZulu-Natal, and is confined to the Pondoland Centre of Endemism (Murray *et al*. 2004). It grows from sea level to 500 m and, according to its authors; it has unique habitat requirements in usually preferring perennially wet, swampy habitats or damp seepages on rock ledges. However, these authors concede that the closely related *C. gardenii* 'does occasionally grow along stream banks or in wetter than usual habitats'. *C. robusta* occurs in small to large colonies usually associated with marshy areas in grassland.

Ernst van Jaarsveld

45

Buttress roots at the base of the pseudostem act as support for larger specimens growing in these marshy conditions, consisting of rotting leaf litter and humus. The plants sometimes occur along stream banks as well as in seepage areas on cliff faces, or in relatively drier rocky terrain adjacent to marshy areas (Murray *et al*. 2004). However, not all *C. robusta* plants are massive, and considerable variation exists from one population to another (and even within the same population) in degree of exsertion of the stamens beyond the perianth, in many instances approaching that of *C. gardenii*. Its geographical separation from *C. gardenii* and specific ecological niche suggests that it could be more appropriately regarded as a subspecies of *C. gardenii*. *C. robusta* is sympatric with *C. miniata* in parts of its range, such as at Oribi Gorge in southern KwaZulu-Natal. The distribution of the closely related *C. gardenii* var. *gardenii* is in central, eastern and northern KwaZulu-Natal, extending north-west and north-eastwards from the Durban area, terminating in northern KwaZulu-Natal (Swanevelder *et al*. 2005). It grows in colonies mainly on the floor of evergreen forest or within the shade of large boulders, or occasionally in the forks of trees, in rotting leaf litter in light to heavy shade. Its occasional presence in trees suggests that birds are likely vectors of the ripe seeds, and in KwaZulu-Natal, bulbuls are known to disperse the seeds (Thurston 1998).

A natural hybrid between *C. miniata* and *C. gardenii* var. *gardenii* was collected by Sean Chubb at Stanger in KwaZulu-Natal in 2003 (Chubb pers. comm.) (see page 105). The rare, pale yellow or lemon-

C. robusta (left), is considered by some authorities to be a separate species, but by others merely as a robust form of *C. gardenii* (right)

Opposite: The remarkably thick mats of fleshy roots of *C. mirabilis* establish themselves in cool niches underneath and between large sandstone boulders

yellow variant of this species, *C. gardenii* var. *citrina* is confined to two disjunct populations, one at the southern extreme near Durban and another at the northern end of the distribution range in northern KwaZulu-Natal (Swanevelder *et al*. 2005).

The most northerly occurring species, *C. caulescens* is found in northern and central Limpopo, northern and eastern Mpumalanga and northern Swaziland. It is mostly lithophytic, occurring on steep sandstone rock faces in semi-shade or sometimes in full sun, clinging to moss-, lichen- and litter-covered rocks in deep shade on margins of evergreen forest, or occasionally as an epiphyte on tree trunks and branches, or on decomposing tree stumps, at altitudes from 1500–1770 m. The ripe berries are transported by samango and vervet monkeys, as well as by the Knysna touraco (Winter 2005a).

Populations on the escarpment in eastern Mpumalanga and northern Swaziland are occasionally subject to very cold conditions with light snowfalls. On the border between northern Swaziland and eastern Mpumalanga it is sympatric with the most northerly populations of *C. miniata,* where a natural hybrid between these two species occurs, *Clivia* x *nimbicola* (see page 103).

C. mirabilis, the most westerly species, is restricted to a few populations on the Bokkeveld Escarpment in the south-western part of the Northern Cape, and a mountainous area in the north-western part of the Western Cape. Occurring in the winter rainfall zone, it grows under relatively dry conditions, experiencing a Mediterranean climate where cool, moist winters with occasional frosts alternate with long, hot, dry summers where temperatures can at times soar to 45 °C. It is thought to be a survivor from a previously more moist climatic period in which a subtropical regime prevailed over what is now a semi-arid region. Its remarkably thick mats of fleshy roots establish themselves in the relatively cool niches underneath and between large sandstone boulders, in a similar manner to those of *Agapanthus africanus* in the mountains of the south-western Cape. *C. mirabilis* occurs in relict afromontane forest within the protection of large shrubs and trees including *Halleria lucida, Olea europaea* subsp. *africana, Maytenus acuminata* and *Podocarpus elongatus* and is found in sheltered valleys, mainly in dappled shade, growing in decaying leaf litter. Plants are occasionally found in positions exposed to direct sun for several hours, and their leaves are clearly under considerably more stress than those found in shade. It is separated from the nearest *C. nobilis* populations in the southern part of the Eastern Cape by some 800 km (Rourke 2002a, b). In attempting to understand why clivias are absent from apparently ideal habitat in forests of the south-western and southern Cape, Snijman (2003) proposed that the natural incidence of fire in the forests of this region and the inability of clivias to survive fires has resulted in the distribution of *Clivia* as we know it today.

Below: *C. mirabilis* occurs in relict afromontane forest within the protection of large shrubs and trees

Opposite: *C. miniata* is the species most widely used by traditional healers

Medicinal uses and conservation

Five of the six *Clivia* species (*C. caulescens, C. gardenii, C. miniata, C. nobilis* and *C. robusta*) are used medicinally by traditional healers of some of the indigenous peoples of South Africa, with *C. miniata* the most widely used, followed by *C. nobilis* (Crouch *et al.* 2003, Swanevelder 2003). They are used for both spiritual and physiological ailments. A large number of isoquinoline alkaloids have been isolated from *C. miniata* and *C. nobilis*, some of which are also found in *C. caulescens* and *C. gardenii*. The highly toxic nature of these alkaloids is evidence of their effectiveness in prepared extracts against viral disease like measles and the Coxsackie virus (Hutchings *et al.* 1996). The plant parts used are the roots, the rhizome and the leaves.

An infusion of the whole plant is used by the Zulu people to sprinkle yards to ward off evil of any description, and of the roots as a love-charm emetic. Chopped *Clivia* parts are used in mixes in combination with a number of other plant species. Both dry and fresh plant mixes are used. In the dried mix, the ingredients are ground to ash, then placed in the mouth and blown towards the sun, while simultaneously the name of the enemy is called out so that evil may befall the enemy. For the fresh mix, fresh ingredients are ground and mixed into a decoction that is then smeared onto the body as a washing agent, in order that it may provide protection from evil curses (Crouch *et al.* 2003). In addition, decoctions of the rhizome are used as a treatment for snakebite and feverish conditions by the Zulus.

In the Eastern Cape, the Xhosa people use decoctions of the roots of *C. miniata* to treat urinary conditions and childlessness. It is also said to relieve pain, while leaf infusions are used as a cleansing agent. Decoctions of the whole plant (except the flowers) are used to ease and hasten childbirth. It should be assumed that all parts of the *C. miniata* plant may cause at least mild stomach upset if ingested, and it is potentially fatal in large doses. The sap of *C. caulescens*, *C. gardenii*, *C. miniata* and *C. nobilis* is known to irritate the skin (Hutchings *et al*. 1996), this no doubt true of *C. mirabilis* and *C. robusta* as well.

All six *Clivia* species are currently under varying degrees of direct threat in the wild as a result of loss of habitat, illegal collecting by traditional healers for sale at 'muthi' markets, and illegal collecting by unscrupulous plant collectors, in that order. Afromontane forest, the most important habitat type of the five summer rainfall *Clivia* species, enjoys scant official protection in South Africa and Swaziland, and is disappearing at an alarming rate, mainly as a result of removal for fuel, agricultural purposes and urbanization (Chubb 1996, Duncan 1999).

The International Union for the Conservation of Nature and Natural Resources (IUCN) have plants listed on their 'Red List' under categories that indicate varying degrees of their probability or risk of extinction in the wild. There are nine IUCN categories under which every species can be classified: 'Extinct' (EX), 'Extinct in the wild' (EW), 'Critically Endangered' (CR), 'Endangered'

Below: Indications are that *C. nobilis* numbers are in continuing decline as a result of 'muthi' collecting

(EN), 'Vulnerable' (VU), 'Near Threatened' (NT), 'Least Concern' (LC) and 'Data Deficient' (DD) (IUCN 2006).

Based on the latest preliminary evaluations of these categories for *Clivia* (Janine Victor, personal communication), *C. caulescens, C. gardenii* and *C. miniata* var. *miniata* are currently listed as 'Least Concern', meaning that they have been evaluated and do not qualify for any of the categories listed above, but may do so in the future. *C. mirabilis, C. nobilis* and *C. robusta* are listed as 'Vulnerable', meaning that evidence indicates that they are facing a high risk of extinction in the wild. Although *C. caulescens*, *C. gardenii* and *C. miniata* var. *miniata* numbers are known to be in decline, they do not meet the criteria necessary to qualify for Red Listing, but in terms of the 'Orange List', an alternative list that caters for taxa worthy of consideration for conservation (Victor & Keith 2004), they are currently regarded as 'Declining'. Immediate identification of clivias at species level in the 'muthi' markets is generally difficult, as typically the inflorescences and tops of the leaves are cut off. *Clivia nobilis*, however, can be quite easily recognized by the characteristic leathery texture of the leaves, and by their minutely serrated margins, and it is considered the most threatened of the *Clivia* species (Snijman & Victor 2003; Victor, personal communication, 2007). Indications are that its numbers are in continuing decline and populations have been severely damaged at Zuurberg, west of Grahamstown, as a result of 'muthi' collecting. Only those found in dense stands in especially rugged terrain appear to be fairly safe from 'muthi' collectors (Cowley 2006b).

Sold in large numbers, *C. miniata* var. *miniata* is widely traded and at great risk of being over-harvested as its numbers are reported to be dwindling increasingly in regional medicinal plant markets, and accessible colonies are becoming increasingly decimated. *C. gardenii* var. *gardenii* is probably the second most popular species in trade, and *C. caulescens* has been noted in markets on the Witwatersrand (Williams 2005). Inhospitable marshy habitat does not appear to be a deterrent against collecting by traditional healers and illegal collectors, as has been noted in some populations of *Clivia robusta* var. *robusta*. This species is particularly vulnerable considering its restricted geographical distribution, its few, severely fragmented populations, and its specific ecological niches (Swanevelder 2003). It occurs in a number of protected reserves throughout its range, but the populations are very localized due to their specific swamp habitat requirement. Fortunately, the yellow form of this species (*C. robusta* var. *citrina*) is protected by the owner of the land on which the only population is known to occur (Swanevelder *et al.* 2006).

Although not traded medicinally, the rather small known populations of *C. mirabilis* are in need of continued maximum protection, in order to prevent exploitation by hobbyist collectors. To this end, a project established by the Northern Cape Departments of Agriculture and Nature Conservation, has provided seedlings raised from wild-collected seed to the public. The project has eased the pressure on the wild populations inside the Oorlogskloof Nature Reserve, and provided employment and upliftment to the local community at Nieuwoudtville. John Winter, former

Curator of Kirstenbosch National Botanical Garden, undertook to support the project by germinating the seeds, growing-on seedlings for sale to the public, and training staff of the Nieuwoudtville Bulb Conservation Project in these methods. Seeds were harvested in the wild in 2003 and the first batch of seedlings was made available to the public via the Internet in 2005. It is estimated there are now almost twice the number of plants in cultivation than in the wild, although it should be borne in mind that a lot has still to be learned regarding the successful cultivation of this plant from seedling to flowering stage. The genus *Clivia* is protected in all nine provinces of South Africa, and as of 1 February 2008, *C. mirabilis* became protected at national level, requiring a possession permit to own or possess material of this species (Faber 2007).

Opposite: Light orange form of *Clivia miniata* in the Camphor Avenue at Kirstenbosch

Below: As of 1 February 2008, *C. mirabilis* became protected at national level, requiring a possession permit to own or possess material of this species

Hein Grebe

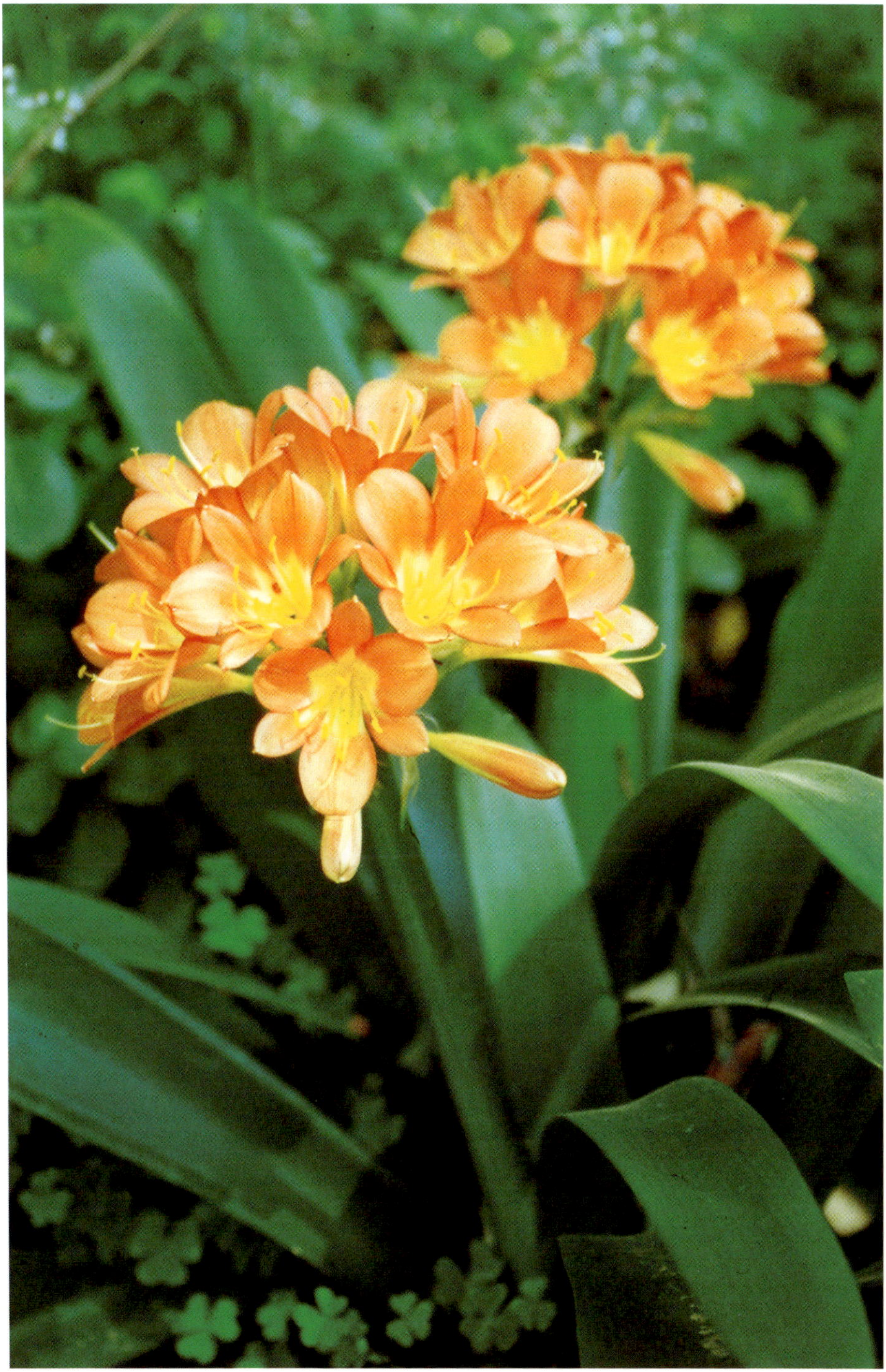

Below: In *C. robusta* (left) the stigma and stamens are usually included within or just emerging beyond the perianth, whereas in *C. gardenii* (right) the stigma and stamens are usually well exserted beyond the perianth

Opposite: A red form of *C. miniata* of cultivated origin

TAXONOMY

Generic description

Clivia Lindl. in Edwards's Botanical Register 14: t. 1182 (1828); Baker in Flora Capensis 6: 228 (1896); Dyer, R.A., The genera of southern African flowering plants 2: 950 (1976); Vorster & Smith in Flowering Plants of Africa 53: 70 (1994); Duncan, G.D., Grow Clivias (1999); Snijman, D.A. in Strelitzia 10: 572 (2000); Koopowitz, H., Clivias (2002). *Imatophyllum* Hook. in Curtis's Botanical Magazine 55 t. 2854 (1828); *Imantophyllum* Hook. in Curtis's Botanical Magazine 80 t. 4783 (1854).

Type species: *C. nobilis* Lindl.

Evergreen, long-lived, summer- or winter-growing, clump-forming or solitary perennials. *Rhizome* vertical, short to long, with a thick mat of perennial, horizontally spreading roots covered with a corky velamen layer. *Leaves* 7–26, pale to dark green or olive green, sheathing, produced basally or atop an aerial stem, distichous, strap-shaped, soft to leathery, flat to strongly canaliculate, spreading to arching or erect, under- or overtopping the inflorescence; upper surface plain or with a dull to prominent white median striation; apex acute to obtuse-apiculate, rounded or notched; margins smooth or serrated. *Inflorescence* a sparse to dense pseudo-umbel produced atop a stout, solid, laterally compressed, 2-edged scape, subtended by 4–7 lanceolate, papyraceous or membranous spathe bracts; bracteoloes linear, papery or membranous; pedicels drooping to erect, slender to thick. *Flowers* trumpet-shaped or tubular, actinomorphic or slightly to strongly zygomorphic, spreading to pendent or erect, unscented or sometimes sweetly fragrant; perianth tube short, tepals overlapping along entire length or flaring above and slightly to well recurved, 3 outer tepals narrower than 3 inner tepals. *Stamens* arising from throat of perianth tube, included or well exserted beyond perianth; filaments linear, straight to slightly curved or spreading, anthers oblong, dorsifixed and versatile. *Ovary* inferior, globose or ovoid, 3–9 ovules per locule, style linear, central, included to well exserted, straight to weakly declinate with a stout to soft, tricuspidate stigma. *Fruit* a globose, subglobose, ovoid or irregularly oblong, red, reddish-pink or yellow berry with pale yellow flesh. *Seeds* globose to ovoid or irregularly shaped, water-rich, brownish-cream, 1–25 per berry, embryo cream-coloured. *Chromosome number*: $2n = 2x = 22$ (Gouws 1949, Ran *et al.* 1999, Zonneveld 2002).

Key to the species

1 Flowers trumpet-shaped, spreading to erect, pedicels thick, radiating . . . . *C. miniata*

1' Flowers tubular, nodding to pendent, pedicels narrow, erect to arched

2 Rhizome caulescent with age, plants lithophytic . . . . *C. caulescens*

2' Rhizome acaulescent, plants terrestrial

3 Perianth straight, pedicels orange-red or purple at flowering . . . . *C. mirabilis*

3' Perianth curved, pedicels green at flowering

4 Leaves leathery, apices notched or rounded, margins serrated . . . . *C. nobilis*

4' Leaves soft-textured, apices pointed, margins entire

5 Stigma and stamens usually included within or just emerging beyond perianth, leaves broad (up to 90 mm diam.), plants robust . . . . *C. robusta*

5' Stigma and stamens usually well exserted beyond perianth, leaves relatively narrow (up to 50 mm diam.), plants gracile . . . . *C. gardenii*

Clivia miniata

A red form of *C. miniata* var. *miniata* from the Eastern Cape

Clivia miniata (Lindl.) Regel var. ***miniata*** in Gartenflora 14: 131, t. 434 (1864).

Type South Africa, KwaZulu–Natal, precise locality unknown, *ex hort Messrs Lee* (CGE, holotype).

Synonyms *Vallota* ? *miniata* Lindley in The Gardener's Chronicle 8: 119 (1854); *Imantophyllum* ? *miniatum* W.J.Hooker in Curtis's Botanical Magazine 80: t. 4783 (1854).

Etymology The specific epithet 'miniata' refers to the flame-scarlet flower colour, 'coloured with red lead or cinnabar'.

Common names Clivia, bush lily, September lily, orange lily, St John's lily, boslelie (Afrikaans), Umayime (Zulu).

Distribution South-west of the Kei River at Haga-Haga in the eastern part of the Eastern Cape to the southern, central, eastern and northern parts of KwaZulu-Natal, to western and northern Swaziland, and eastern Mpumalanga. It is sympatric with *C. nobilis* in the Eastern Cape, with *C. robusta* in the Eastern Cape and southern KwaZulu-Natal, with *C. gardenii* in central, eastern and northern KwaZulu-Natal, and with *C. caulescens* in eastern Mpumalanga and northern Swaziland.

Habitat In shelter of cool afromontane forest, in large numbers on ledges and shady ravines, mainly on south-, south-east- and west-facing slopes in coastal and midland forest, between sandstone boulders and on doleritic soils, from just above sea level to 1500 m. Subject to short periods of snow in northern Swaziland and eastern Mpumalanga.

Flowering period Early August to mid November (early spring to early summer), or sporadically at any other time of year.

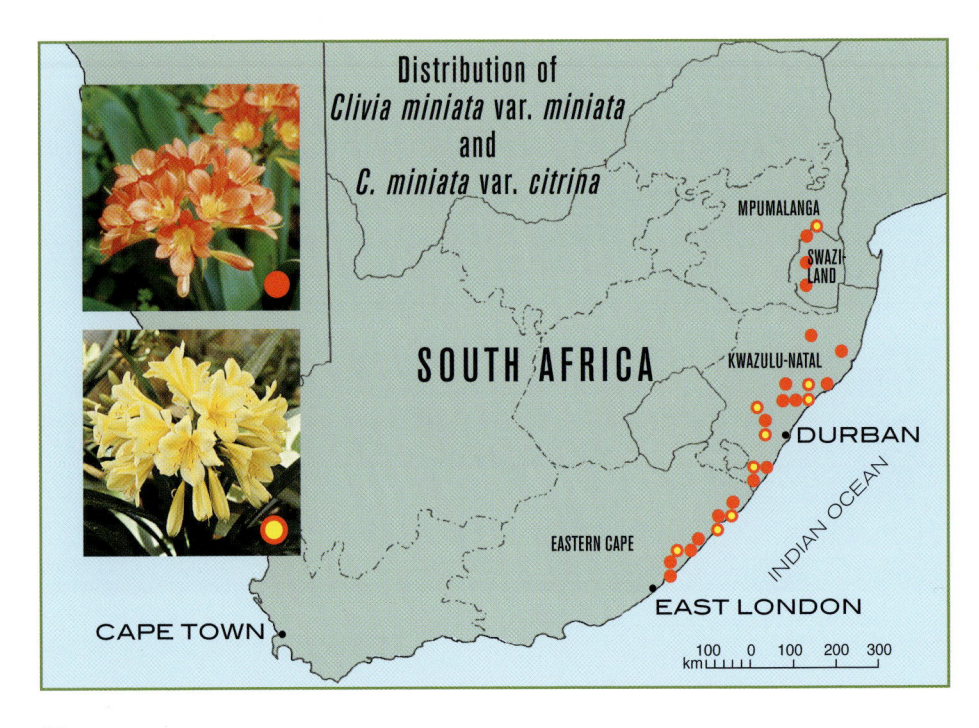

Distribution of *Clivia miniata* var. *miniata* and *C. miniata* var. *citrina*

Description Extremely variable evergreen, summer-growing, offset-forming perennial 0.5–1 m high. *Rhizome* subterranean or rarely shortly caulescent up to 100 mm high with advanced age, with thick fleshy roots up to 10 mm wide. *Leaf sheath* uniformly green, rarely tinged with pale to dark maroon. *Leaves* 12–26, strap-shaped, pale to dark green, sub-erect to arching, strongly distichous, flat to weakly canaliculate, 450–900 x 25–70 mm, apex acute. *Scape* erect to sub-erect, sturdy, narrowing towards apex, pale to dark green, 300–600 x 25–40 mm. *Inflorescence* large, dense, 16–30-flowered, held above the leaves, subtended by 3–5 translucent white, lanceolate, membranous spathe-bracts, 35–60 x 8–15 mm; bracteoles linear, membranous, 10–13, translucent white, 18–30 x 1–3 mm; pedicels sturdy, 3-sided, thick, sub-erect to erect, dark green, 45–90 mm long. *Flowers* trumpet-shaped, spreading to erect; perianth tube funnel-shaped, 7–10 mm long, outer surface pale to dark orange or orange-red, rarely pale peach-pink, inner surface creamy-yellow with dark yellow or green median keel; outer tepals broadly lanceolate, weakly canaliculate, upper half uniformly pale to dark orange, orange-red or rarely pale peach-pink, shading to creamy-yellow in lower half, 38–60 x 15–20 mm, margins flat or weakly undulate, apex apiculate; inner tepals spathulate, weakly canaliculate, protruding slightly beyond outer tepals, straight or slightly to moderately re-curved, upper half pale to dark orange, orange-red or rarely pale peach-pink, shading to creamy-yellow in lower half, 40–65 x 15–25 mm; margins flat or weakly undulate, apex rounded. *Stamens* weakly to strongly declinate, included within perianth, filaments linear, white or pale to dark yellow, arising from throat of perianth tube, 35–48 mm long; anthers oblong, 3–7 x 1–2 mm, pollen yellow at anthesis. *Ovary* globose or oblong, dark green, 6–8 x 5–7 mm, containing 5–9 ovules per locule; style linear, straight to weakly declinate, creamy-white or yellow, included or protruding beyond perianth, 40–70 mm long, stigmatic branches green or yellow, very short, 1.0–1.5 mm long. *Fruit* a globose or sub-globose, glossy dark red berry 20–30 x 17–27 mm, ripening within 9–12 months, from midwinter to early spring. *Seeds* irregularly shaped, pale brown, 10–20 x 10–18 mm, 1–25 per berry.

Cultivation Needing no introduction, *C. miniata* var. *miniata* is horticulturally the most important of all the *Clivia* taxa. This extremely variable plant is one of the most valuable evergreen geophytes available to gardeners, and is the variety most commonly grown in parks, gardens and conservatories, and as an indoor plant. Its ease of culture in a range of growing media and its reliable flowering performance has secured its place both as a valuable landscape subject in temperate climates, as well as a highly successful container plant in colder parts. Flower colour predominates in shades of orange and orange-red, but pastel-flowered forms are still rare in cultivation.

Thriving in dappled shade, it is remarkably drought tolerant during its winter rest phase, yet easily withstands heavy winter rainfall over this period, provided the growing medium is sufficiently well draining. It has a long flowering period extending from late winter to early summer and makes an admirable cut-flower. It is a gross feeder, benefitting greatly from applications of organic fertilizers like Neutrog Bounce Back to the soil surface or incorporated into the growing medium. Constriction of the roots encourages flowering and the plants like to remain undisturbed for many years. It does not perform well in overly humid conditions prevalent in summer in the Lowveld region of Mpumalanga and the coastal parts of KwaZulu-Natal, or the excessively hot, inland parts of the Great Karoo.

Most forms of this variety multiply readily by offset formation, and under ideal conditions, seedlings can usually be brought into flower in their third spring season. The foliage is highly susceptible to attack by lily borer caterpillars (see page 159) and snout beetles (see page 162) in southern Africa, as well as to a number of fungal diseases (see page 164) and to bacterial soft rot (see page 167). In addition, European garden snails cause great damage to scapes, developing flower buds and open flowers (see page 162). The form from Mbotye in the Eastern Cape is one of the earliest to flower (early August) while those from The Bearded Man Mountain on the boder between Swaziland and South Africa are amongst the latest (mid-October to November).

Due to their rarity both in the wild and in cultivation, pastel-flowered forms of *C. miniata* var. *miniata* tend to be assigned cultivar names more readily than orange and red-flowered forms. Examples of some superior South African pastel-flowered forms include:

Clivia miniata var. *miniata* 'Appleblossom'

Collected in the former Transkei, Eastern Cape, by John Winter in the late 1990s, this is a truly outstanding form with white flowers delicately suffused with pale peach on the upper inner surface of the three outer tepals and at the tips of the three inner tepals. The flowers are relatively widely spaced and the inflorescence is borne at approximately the same level as the tips of the leaves. The tepals overlap for more than half their length, the tips of the inner and outer tepals are slightly re-curved and the bright yellow anthers provide excellent contrast. The flower colour of this clone (also known as Q2) is so similar to those of apple blossoms that it was named 'Appleblossom' by John Winter. It is one of seven plants collected at the same locality, each of which has slightly different colouration (see next page).

Clivia miniata var. *miniata* 'Chubb's Peach'

The first plant was found in the Ngwahumbe River Valley near Eston, KwaZulu-Natal in the 1950s (Thurston 1998). The well balanced, spherical inflorescence is carried above the leaves and the deep peach tepals are free almost to their bases, with slightly undulate margins. The relatively large flowers are scented and tepal colour darkens as the flower matures. The plants multiply readily by offset formation (see page 62).

Clivia miniata var. *miniata* 'Mbashe Peach'

The original plant was collected by John Winter at Mbashe in the Transkei, Eastern Cape, in the late 1990s. It is an outstanding form with a fairly dense flower head held well above the leaves, consisting of relatively large, widely flaring blooms, the tepals flushed dark peach in the upper half, shading to cream in the lower half, with a yellow throat. The tepals overlap for approximately half their length and their margins are distinctly undulate, the prominent white style protruding well past the perianth (see page 63).

Clivia miniata var. *miniata* 'Appleblossom' (see previous page)

Clivia miniata var. *miniata* 'Chubb's Peach' (see page 60)

Clivia miniata var. _miniata_ 'Naude's Peach' (not illustrated)

Of unknown origin, this magnificent form has a dense, well rounded inflorescence of relatively small flowers, held well above the leaves. The relatively broad tepals are pale peachy-yellow on opening, maturing to a clear, dark peach, and are strongly overlapping for about three quarters of their length. It is a vigorous plant, often producing two inflorescences per year, but is slow to produce offsets.

Clivia miniata var. _miniata_ 'Ndwedwe Gamma Peach' (not illustrated)

Discovered as recently as 1996 at Ndwedwe in Upper Tongaat, KwaZulu-Natal, from an area that has also yielded two outstanding clones of *C. miniata* var. *citrina* (see page 67). An outstanding feature of the flowers is that the tepals are free all the way to the base; they are relatively narrow with distinctly rounded tips and upon opening are pale peach with a pale green median keel that matures to dark peach, the edges of the tepals becoming peachy-white towards the end of the flowering period. It is a robust form, usually producing two inflorescences per season, but appears not to produce offsets readily.

Clivia miniata var. *miniata* 'Mbashe Peach' (see page 60)

Clivia miniata (Lindl.) Regel var. ***citrina*** Watson in The Gardener's Chronicle (3rd series) 25: 228 (1899).

Type South Africa, KwaZulu–Natal, precise locality unknown (cultivated in garden of Mrs Powys Rogers in Cornwall, England), figure by H.G. Moon in The Garden, plate 1246 (1899).

Synonym *Clivia miniata* (Lindl.) Regel var. *flava* E.Phillips in The Flowering Plants of South Africa 11: t. 411 (1931).

Etymology The specific epithet 'citrina' refers to the pale yellow flowers.

Common names Yellow clivia, yellow bush lily.

Distribution Occasional in the eastern part of the Eastern Cape, southern, central and eastern KwaZulu-Natal, northern Swaziland and eastern Mpumalanga.

Habitat As for *C. miniata* var. *miniata* (see page 58).

Flowering period August to November, or sporadically at any other time of year.

Description As for *C. miniata* var. *miniata*, but flowers are pale to deep yellow, occasionally with greenish throats, and not pale to deep orange, orange-red, orange-pink or peach pink as in var. *miniata*. Fruits are pale to deep yellow, sometimes with reddish speckling, not dark red as in var. *miniata,* ripening from midwinter to early spring. Most forms of var. *citrina* tend to be somewhat shorter than var. *miniata*, seldom reaching more than 0.8 m.

Cultivation As for *C. miniata* var. *miniata*, but the foliage is generally more susceptible to fungal disease, especially the *Agapanthus* fungus, leaf spots and rust fungi (see page 164, 165). In a garden setting, the var. *citrina* is very effectively displayed when interplanted with orange and red-flowered forms of var. *miniata* that flower simultaneously.

Since the discovery of the first plants of *C. miniata* var. *citrina* in the wild at Eshowe in Zululand in about 1888, numerous different forms of this variety have been found in various parts of the Eastern Cape and KwaZulu-Natal, and, to a lesser extent, in Mpumalanga and Swaziland. Forms of var. *citrina* have also arisen spontaneously in cultivation, and many more artificial hybrids between different forms of the var. *citrina* have been raised in various parts of the world; some of these are true-breeding yellows, while others are not. Unfortunately, in many instances the exact origin and parentage of these naturally occurring forms and hybrids in cultivation is unknown. The forms that breed true to type are classed as 'Group 1', while those that do not are classed as 'Group 2 '(Morris 1999). In yellow-flowered forms belonging to 'Group 1', there is no trace of anthocyanin pigment in any part of the plant, including the fruit, and they breed true to type when self-pollinated (selfed), or inter-crossed with other forms belonging to 'Group 1'. In 'Group 2 ', the flowers have traces of anthocyanin (pinkish-red) pigment streaks or spots in their flowers and fruits. When a yellow-flowered individual belonging to 'Group 2' is crossed with another yellow-flowered individual belonging to 'Group 2', the progeny is usually 100% yellow, but when a 'Group 1' is crossed with a 'Group 2' (irrespective of which is the pod parent or the pollen parent), the progeny is usually 100% orange. When selfed, the progeny of yellows belonging to 'Group 2' varies from 100% orange to 50% yellow, or a very low percentage of yellow (for example 'Dwesa Yellow' produces a few yellows when selfed).

Some excellent South African forms of *C. miniata* var. *citrina* include:

Clivia miniata var. *citrina* 'Butter Yellow' (see page 66)

Clivia miniata* var. *citrina 'Butter Yellow' (Group 2)

A single plant arose spontaneously within a planting of *C. miniata* var. *miniata* at 'The Wilds' in Johannesburg. It was donated to the Kirstenbosch bulb collection by Rodney Saunders in 1988, who gave it its cultivar name. It is a floriferous plant, its rich yellow flowers with their deeper yellow throats are produced in a well-rounded, full flower head. The lower half of the somewhat narrow tepals is distinctly channelled and their upper halves are non-overlapping and flared, giving them an almost flat appearance. This clone was used as the pod parent in a cross with 'Natal Yellow', a selection of which was named 'Pat's Gold' (see page 65).

Clivia miniata* var. *citrina 'Dwesa Yellow' (Group 2) (see next page)

Synonym 'Transkei Yellow'

Found near the Mbashe River Mouth in the Transkei, Eastern Cape, in the late 1930s, the plant produces relatively small, pale cream flowers with slightly deeper throats in a loose flower head. The tepals overlap in the lower half and the stamens are included within the perianth.

Clivia miniata* var. *citrina 'Kirstenbosch Yellow' (Group 1)

Of unknown origin, two plants were obtained in 1951 by Kirstenbosch from the erstwhile Reed's Nursery in Wynberg, Cape Town (see page 68). By 1979 they had increased vegetatively to just five individuals. It is a large-flowered, slow-growing, long-lived form with a strong fragrance reminiscent of azaleas (Duncan 1985). Producing a large, rounded inflorescence, the relatively narrow tepals are well reflexed and the plants have unusually broad, light green leaves that are highly susceptible to fungal attack, causing them to die back from the tips and requiring a rigorous spraying programme. The plant forms a short, aerial stem with advanced age. Seedlings of 'Kirstenbosch Yellow' are 100% true to type when selfed, and the fragrance of this clone is maternally inherited when used in hybridization. A cross made by the author at Kirstenbosch using 'Kirstenbosch Yellow' as pod parent and 'Natal Yellow' as pollen parent resulted in 100% orange progeny, one of which was selected and named 'Kirstenbosch Supreme' (see page 116).

Clivia miniata* var. *citrina 'Mare's Yellow' (Group 1) (not illustrated)

Synonym 'Howick Yellow'

Collected from the base of the Howick Falls in KwaZulu-Natal in the early 1890s, the flowers are a clear pale yellow, and the rather narrow tepals overlap in the lower half and have darker yellow median keels. The plant readily produces suckers and has narrow, dark green leaves.

Clivia miniata* var. *citrina 'Mvuma Yellow' (Group 2) (not illustrated)

Collected as recently as the late 1970s on the farm 'Mvuma' near the Tongaat River in KwaZulu-Natal, it produces most attractive rich yellow flowers with deeper yellow median keels. The relatively broad tepals overlap for more than half their length, and the inner tepal apices are distinctly rounded. The inflorescence is rather sparse.

Clivia miniata* var. *citrina 'Natal Yellow' (Group 2)

Collected at Bainesfield near Richmond in KwaZulu-Natal by Cynthia Giddy, it has one of the richest flowers of the natural

Clivia miniata var. *citrina* 'Dwesa Yellow' (see previous page)

yellows, and the flower buds and newly opened tepals are distinctly tinged with green. An exceptionally vigorous plant, it has narrow, dark green leaves and reproduces rapidly by numerous stolons produced from the top of the rhizome, each resulting in a new plant. The medium sized flower head is well rounded and the relatively broad tepals overlap for more than half their length and are slightly recurved. The flowers are unscented and self-sterile, only producing yellow offspring when crossed with another Group 2 plant. The plant flowers late in the season and the fruits are bright yellow with tiny red speckles. A single plant was obtained from the late Cynthia Giddy in September 1981 (Duncan 1985). It was used as a pod parent for a cross with 'Naude's Peach', in which all seedlings were pigmented, but the flowers produced the most interesting pastels with large white, yellow or green throats that develop a pink blush as they mature (see page 69).

Clivia miniata var. ***citrina*** 'Ndwedwe Alpha Thurston' (Group 2)

Collected at Ndwedwe in Upper Tongaat in KwaZulu-Natal in the mid 1980s, this is a robust and exceptionally beautiful plant with a large, rounded, fairly dense umbel of clear yellow, slightly scented blooms. The tepals overlap for half their length and the stamens are included within the perianth. Another yellow-flowered clone from the same area at Ndwedwe is 'Ndwedwe Beta Thurston' that produces a smaller, less rounded flower head with fewer flowers and slightly exserted stamens (see page 69).

Clivia miniata var. *citrina* 'Kirstenbosch Yellow' (see page 66)

Clivia miniata var. *citrina* 'Natal Yellow' (see page 66)

Clivia miniata var. *citrina* 'Ndwedwe Alpha Thurston' (see page 67)

Clivia miniata var. ***citrina*** 'Noyce's Sunburst' (Group 1)

Indications point to this form originating from Eshowe, although these have not been entirely verified. Producing a magnificent globe-like flower head, it is a broad-leafed plant with a strong fragrance reminiscent of azaleas. It is similar to 'Kirstenbosch Yellow,' differing mainly in its larger overall size, with the inflorescence produced well above the leaves on a broad, robust scape (Duncan 1999). Its berries produce a larger crop of seeds than 'Kirstenbosch Yellow' but it is a similarly slow-growing form that does not flower reliably every year. Like those of 'Kirstenbosch Yellow', the plant produces a short aerial stem with advanced age. Material was donated to Kirstenbosch in February 1993 by Michael Noyce of Kloof, KwaZulu-Natal, a past member of the Board of Trustees of the National Botanical Institute.

Clivia miniata var. *citrina* 'Noyce's Sunburst'

Clivia caulescens

Clivia caulescens from north-eastern Mpumalanga

Clivia caulescens R.A.Dyer in The Flowering Plants of South Africa 23 t. 891 (1943).

Type South Africa, Mpumalanga, 2430 (Pilgrim's Rest): MacMac forest margin (–DD), *F.Z. van der Merwe s.n.* (PRE, holotype).

Etymology The specific epithet 'caulescens' refers to the distinct aerial stem that develops in mature plants.

Common name Stem clivia.

Distribution Isolated large populations in northern Swaziland, eastern and northern Mpumalanga, and central and northern Limpopo. Sympatric with *C. miniata* in northern Swaziland and eastern Mpumalanga.

Habitat Ravines on bare sandstone rock in afromontane forest, in full sun or amongst shrubs, usually lithophytic, the roots clinging to moss, lichen and leaf mould on rotting tree trunks, sometimes epiphytic in tree forks, the seeds having been deposited there by birds or monkeys, frequently on south- and south-east facing slopes, subject to regular mist, at altitudes from 1500–1770 m, occasionally experiencing snow for short periods.

Flowering period Mainly early September to early November (spring to early summer), or sporadically during the year.

Description Evergreen, summer-growing, lithophytic or rarely epithytic, solitary or offset-forming perennial 0.5–2 m high. *Rhizome* up to 60 mm wide, extending above ground up to 1 m long or more, with thick fleshy roots up to 8 mm wide. *Leaf sheath* uniformly green, sometimes greenish-red. *Leaves* 11–14, strap-shaped, flat to weakly canaliculate, bright

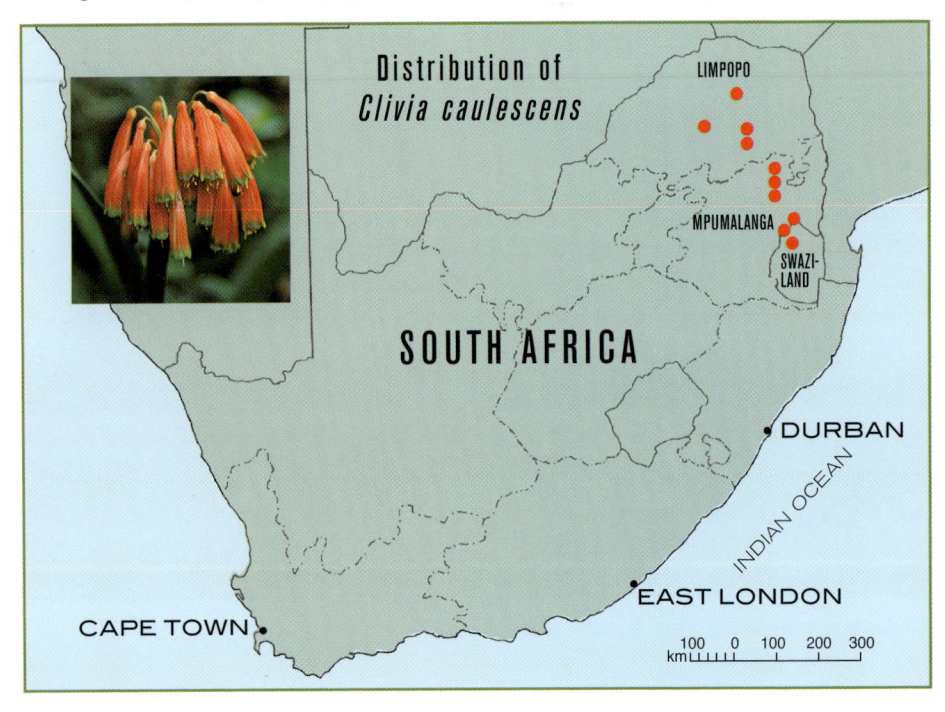

Distribution of *Clivia caulescens*

Clivia caulescens from God's Window, north-eastern Mpumalanga

green to dark olive-green, sub-erect to arching, 320–750 x 30–60 mm, apex acute or obtuse-acute. *Scape* erect, dark green to purplish-maroon, narrowing towards apex, 300–400 x 35–40 mm. *Inflorescence* sparse to dense, flattened on one side, 15–50-flowered, held well above leaves, subtended by 3–4 lanceolate, papery spathe bracts, 40–50 x 10–15 mm; bracteoles 7–10, linear, papery, 20–35 x 2–3 mm; pedicels arched at peak flowering, becoming sub-erect to erect at early fruiting, then arched at late fruiting, green to purplish-green or orange-red, 13–30 mm long. *Flowers* tubular, slightly curved, drooping; perianth tube narrow, 6–8 mm long, pale orange to deep orange-red, rarely pale pink or pale yellow; outer tepals linear, canaliculate, 24–26 x 3–4 mm, pale orange to deep orange-red, rarely pale pink or pale yellow, with slightly flared, pale to bright green tips; inner tepals oblanceolate, canaliculate, wider and protruding slightly beyond outer tepals, uniformly pale orange to deep orange-red, or translucent white with a broad, pale to dark orange median keel, rarely pale pink or pale yellow, with flared, pale to bright green tips, 26–28 x 6–8 mm. *Stamens* slightly curved, retained within perianth or slightly exserted; filaments linear, white, 25–27 mm long, anthers oblong, 3 x 1 mm, pollen yellow at anthesis. *Ovary* globose, pale orange to deep orange-red, 4–5 x 4–6 mm, containing 3–4 ovules per locule; style linear, straight, white, 34–36 mm long, stigma with tricuspidate apex. *Fruit* a globose, dark red berry 15–20 x 15–20 mm, ripening within 6–8 months, from mid- to late winter. *Seeds* pale brown, globose or irregularly shaped, 10–13 x 10–13 mm, 1–5 per berry.

Cultivation *C. caulescens* is a most rewarding and long-lived plant in cultivation, whether grown as an indoor specimen plant, on a protected shady patio, or out in the garden. Its most characteristic feature is the curious aerial stem that develops with advanced age and provides added interest. Under ideal conditions plants flower reliably every year. Certain forms are relatively fast growing and develop an aerial stem much sooner than others. The rhizome and roots are highly susceptible to over-watering and the plant likes an extremely well drained medium such as equal parts of composted pine bark and coarse river sand, with the addition of the organic fertilizer 'Neutrog Bounce Back' that can be mixed into the medium or sprinkled onto the surface. Well established plants are remarkably drought tolerant (provided they are grown in sufficient shade) and during winter the plants enter a dormant phase, when minimal watering is required, although they are able to survive heavy winter rainfall (such as is experienced in the southern suburbs of Cape Town) quite easily provided excellent drainage is provided. During active growth in summer, a heavy drench once per week is sufficient, ensuring that the medium dries out thoroughly before the next application is given.

The plants readily produce offsets, and seedlings take a minimum of three to four years to flower for the first time; restriction of the roots through underpotting encourages flowering. In addition to their main flowering period in spring and early summer, an additional flush of blooms is often produced in autumn, but sporadic blooms may appear at any time of year. The foliage and peduncle are highly susceptible to attack by lily borer caterpillars in summer.

Clivia mirabilis from Oorlogskloof, Nieuwoudtville

Clivia mirabilis Rourke in Bothalia 32,1: 1–7 (2002).

Type South Africa, Northern Cape, 3119 (Calvinia): Nieuwoudtville, Oorlogskloof Nature Reserve, (–AC), *Rourke 2220* (NBG, holotype, BOL, K, MO, NSW, PRE, isotypes).

Etymology The specific epithet 'mirabilis' refers to the miraculous discovery of this most unusual species in seemingly inhospitable habitat for a member of this genus.

Common name Miracle clivia; Oorlogskloof bush lily.

Distribution Bokkeveld Plateau in the south-western part of the Northern Cape near Nieuwoudtville, and mountainous area in the north-western part of the Western Cape near Vanrhynsdorp.

Habitat Humus in cracks between large sandstone boulders, on east-facing cliffs, at 850–900 m above sea level. Occurs in dappled shade of relict afromontane forest. Plants have mats of very thick roots that survive under sandstone rock slabs.

Flowering period Late October to late November (late spring to early summer).

Description Evergreen, winter-growing, solitary (or occasionally offset-forming in cultivation) perennial 0.6–1.2 m high. *Rhizome* short, subterranean, 220–250 mm long, with a large, thick mat of robust fleshy roots 10–20 mm in diam., covered with a thick velamen layer. *Leaf sheath* broad, conspicuously tinged with dark maroon. *Leaves* 8–10, strap-shaped, spreading to sub-erect or erect, blades very leathery, strongly canaliculate, dark green, upper surface with a narrow to relatively broad pale white median striation in younger leaves, gradually fading in older leaves, margins leathery, apex obtuse-acute, 0.6–1.2 m long, 30–70 mm wide.

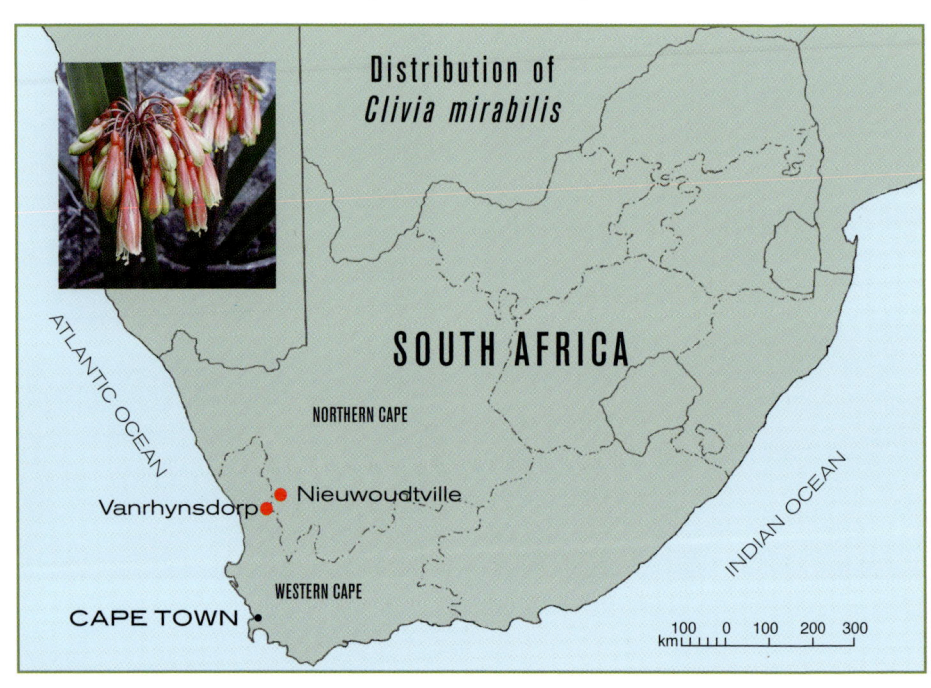

Distribution of *Clivia mirabilis*

SOUTH AFRICA

NORTHERN CAPE

Vanrhynsdorp • Nieuwoudtville

WESTERN CAPE

CAPE TOWN

Above: Coloured pencil drawing of *Clivia mirabilis* by Sibonelo Chiliza, reproduced by courtesy of Prof. Roger Fisher

C. mirabilis in habitat near Vanrhynsdorp (photographs by Hein Grebe)

Clivia mirabilis from Oorlogskloof, Nieuwoudtville

Scape erect to sub-erect, strongly flattened, 300–800 x 9–14 mm, brownish-maroon at flowering, shading to dark green at fruiting. *Inflorescence* sparse to dense, 20–48-flowered, subtended by 5–7 lanceolate to narrowly cymbiform-acute, deep reddish-maroon spathe bracts with green tips, 35–50 x 10–15 mm; pedicels arched, orange-red or purple at flowering, turning green during fruiting stage, 23–40 x 1–1.2 mm. *Flowers* tubular, straight or very slightly curved, flaring towards apex; perianth tube pale to deep orange-red or salmon pink, 10–15 mm long, outer tepals narrowly oblong, orange-red to salmon pink, apices acute, green prior to anthesis, turning yellow at anthesis, 25–42 x 8–12 mm; inner tepals slightly broader, oblanceolate, orange-red to salmon pink with pale green median keels, apices obtuse, green prior to anthesis, turning yellow at anthesis, protruding slightly beyond outer tepals, 27–42 x 10–14 mm. *Stamens* straight or very slightly curved, included or just emerging beyond perianth, filaments linear, white, 25–30 mm long, anthers oblong, pollen yellow at anthesis. *Ovary* ovoid, greenish yellow prior to anthesis, turning bright orange-red at anthesis, 7–8 x 4–5 mm, containing 3–4 ovules per locule; style linear, white, included within perianth at anthesis, becoming shortly exserted at fruiting, 40–45 mm long. *Fruit* a glossy dark red, irregularly oblong to ovoid berry, upper portion narrowed to an obtuse apex, 10–30 x 10–15 mm, ripening within 5 months, from mid to late autumn. *Seeds* ovoid, pale creamy-brown, 12 x 10 mm, 1–3 per berry.

Cultivation *C. mirabilis* is undoubtedly the most difficult of all the *Clivia* species to maintain in cultivation over an extended period. Its roots and base of the pseudostem are extremely sensitive to over-watering, especially those of seedling plants, rapidly rotting off under imperfectly drained conditions. The accumulation of water in the hollow formed in the centre of the plant must be prevented as it can lead to rotting of developing leaves. The plants require an extremely well drained medium, a recommended one being equal parts of finely milled pine bark, milled pine needles and washed, coarse river sand or coarse industrial sand. Care must be taken to ensure that the base of the pseudostem rests at, or just above soil level.

Although the plants are more sun tolerant than the other *Clivia* species and are exposed to full sun for a portion of the day in habitat, they should not be regarded as subjects for very hot, all-day full sun positions under cultivation (Duncan 2004b). A lightly shaded position – a little more sun than that in which *C. nobilis* thrives – receiving 1–2 hours of morning sun followed by semi-shade for the rest of the day suits them well. They are probably able to withstand light frost for short periods. Plants grown in full shade at Kirstenbosch have performed well, but their flower colour tends to be rather washed out. Unlike the five other *Clivia* species, *C. mirabilis* requires a relatively dry summer, which is its natural dormant period. During this time a thorough watering approximately once per month is suggested, increasing to once approximately every ten days to two weeks (depending on climatic conditions) during the active winter growing period. The growing medium should be checked carefully before watering; it should be almost dry before the next watering is given. The plants benefit from annual re-potting into new growing medium to prevent compaction and provide adequate aeration. A plant whose roots have rotted soon falls over. It should then be lifted, cleaned and treated with a fungicide such as captab (e.g. Kaptan) and be planted in sterilized, coarse sand and placed in a warm, shaded position; following which new roots will usually develop. Once seedlings have reached about 400 mm in height, with a pseudostem diameter of about 15 mm, they appear to overcome their susceptibility to rotting. Seedlings perform best with additional heat provided in winter (minimum 15 °C) and a maximum of 30 °C in summer (Winter 2005b). *C. mirabilis* is not nearly as floriferous as the summer-growing species and does not flower readily every year. Experience at Kirstenbosch thus far has been that flowers appear every second or third year. *C. mirabilis* is unable to withstand the rigours of indiscriminate garden watering, and is best suited to cultivation in containers, preferably under cover in areas with heavy winter or summer rainfall. Under ideal conditions, seedlings of *C. mirabilis* are remarkably fast growing in comparison with those of the rather sluggish *C. nobilis,* and will probably reach maturity before the latter species.

C. mirabilis is tolerant of greater extremes of temperature than most forms of the five summer-growing species, being subject to occasional light frosts in winter and scorching summers where daytime temperatures may at times reach 35 °C in the shade.

The leathery leaves of *C. mirabilis*, like those of *C. nobilis*, are too tough to fall prey to the devastating exploits of lily borer caterpillars, or to the nocturnal forays of snout beetles, but a close watch should be kept on mealy bugs that make themselves at home among the leaf bases. Owing to its rather stringent growing conditions, *C. mirabilis* will probably remain a subject for the specialist grower, but its potential in breeding programmes undoubtedly holds great promise for the future. Already, most satisfactory results have been obtained in crosses with *C. miniata* that have flowered in their third season (see page 110) and the dark maroon pigmentation of the leaf bases in *C. mirabilis* is passed on to the progeny of crosses with *C. miniata*. Crosses with *C. nobilis* forms from sunny habitats may even produce drought resistant, fully sun-hardy plants in future.

Clivia nobilis from East London

Clivia nobilis Lindl. in Edwards's Botanical Register 14: t. 1182 (1828).

Type South Africa, Eastern Cape, precise locality unknown (cultivated at Syon House, England), figure by M. Hart in Edwards's Botanical Register, plate 1182 (1828).

Synonym *Imatophyllum aitoni* W.J.Hooker in Curtis's Botanical Magazine 55: t. 2856 (1828).

Etymology The specific epithet 'nobilis' refers to the noble Lady Clive, Charlotte Florentia, Duchess of Northumberland.

Common names Bush lily, boslelie (Afrikaans).

Distribution Confined to the Eastern Cape, mainly in coastal areas, from the Zuurberg Mountains near Grahamstown in the west, to just north of the Kei River in the east, in the Albany Centre of Endemism. Sympatric with *C. miniata* in the northern part of its range.

Habitat Primary and secondary alkaline coastal dunes, and under dense canopy of inland evergreen afromontane forest, on steep, rocky slopes, river banks and forest margins, in acid loamy soil, or in acid leaf litter of the forest floor. Occurs from sea level to 600 m, sometimes semi-lithophytic, growing in leaf debris on large sandstone boulders in forests.

Flowering period Mainly early July to early December (midwinter to early summer) or sporadically at any other time of year.

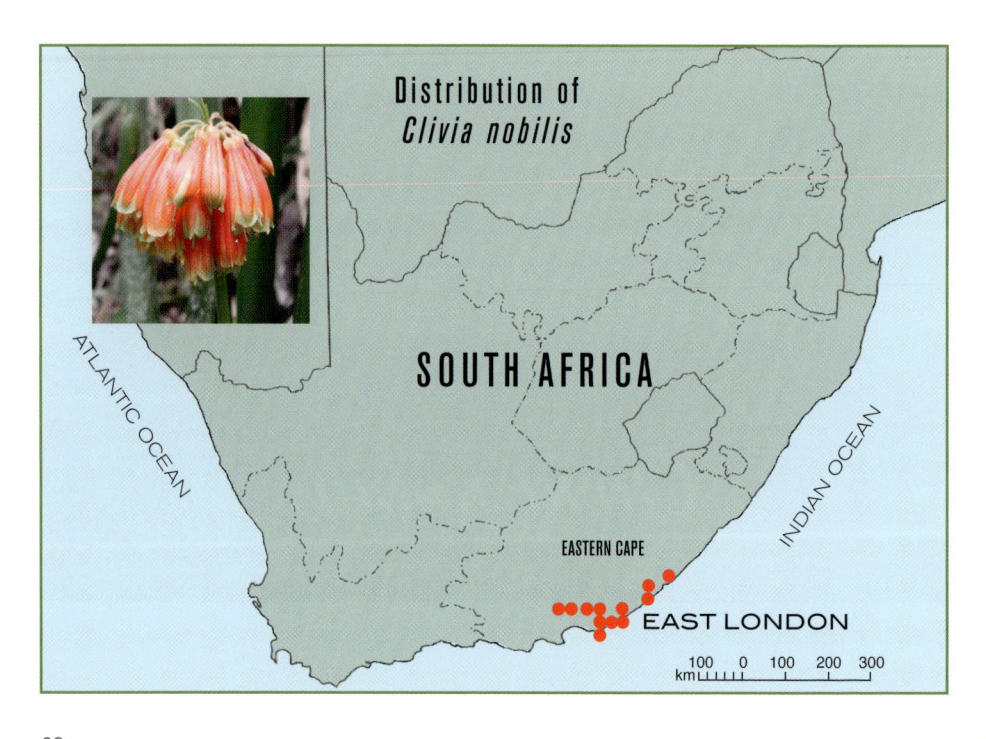

Distribution of *Clivia nobilis*

Description Evergreen, summer-growing, offset-forming perennial 0.5–1.1 m high. *Rhizome* short, subterranean, rarely lithophytic, with a thick mat of spreading fleshy roots. *Leaf sheath* uniformly green, rarely flushed pale to deep maroon. *Leaves* 7–14, leathery, stiffly sub-erect to arched, plain green to glaucous, weakly to strongly canaliculate, 0.3–1 m long, 20–45 mm wide, upper surface plain, rarely with narrow dull white median striation, leaf margins distinctly serrated with a cutting edge, apex rounded or weakly to strongly notched. *Scape* suberect to erect, dark green to brownish-green, 260–440 x 12–25 mm. *Inflorescence* sparse to dense, 20–50-flowered, subtended by 3–6 oblong, translucent, membranous spathe bracts, 22–30 x 5–15 mm; bracteoles filiform, white, 12–18 mm long; pedicels green, arched to spreading, 8–16 mm long; *Flowers* tubular, slightly curved, strongly pendent to drooping, producing copious nectar, perianth very variable in colour; perianth tube pale orange, yellow, orange-pink or deep orange to orange-red, 5–7 mm long; outer tepals oblong, pale orange, yellow, pinkish-orange or deep orange to orange -red, canaliculate, with pale to bright green tips, 18–20 x 4–6 mm; inner tepals narrowly spathulate, canaliculate, protruding slightly beyond outer tepals, midrib pale orange, yellow, orange-pink or deep orange to orange-red, with translucent white sides and pale to bright green tips, 17–19 x 8–10 mm. *Stamens* slightly curved, arising from throat of perianth tube, retained within perianth tube or anthers slightly exserted, filaments sturdy, white, 18–22 mm long, anthers oblong, 2–3 x 1 mm. *Ovary* globose, green to pale yellow, 4–6 x

Below: A red form of *Clivia nobilis*

5–6 mm; style straight to slightly curved, white, slightly exserted, 27–30 mm long. *Fruit* a globose, glossy, bright red, purplish-maroon or occasionally bright yellow berry 15–20 x 15–20 mm, ripening within 6–8 months, in midwinter and spring. *Seeds* pale creamy-brown, globose or irregularly shaped, 7–10 x 7–10 mm, 1–4 per berry.

Cultivation Low-growing forms of *Clivia nobilis* from primary and secondary coastal dunes of the Eastern Cape make outstanding container plants, while larger forms from heavily shaded coastal and inland forests are ideally suited to garden cultivation. The roots of all forms of this species are particularly shallow-seated, allowing the plants to be grown in shallow, wide containers or in shallow, well-drained ground, and the plants benefit greatly from an annual application of well decomposed compost. Once established, the plants are remarkably drought tolerant under sufficiently shaded conditions, and readily produce offsets. Sustained periods of excessive moisture lead to rapid rotting of the roots and experience at Kirstenbosch has shown that watering of container-grown plants in the heat of summer is best carried out by not watering directly onto the leaves or crowns of the plants but rather in a circle, some distance away from the plants.

Seedlings of *C. nobilis* are notoriously slow to develop compared with the other five species, an observation made as far back as the early nineteenth century by the Rev. William Herbert in his classic work *Amaryllidaceae* (Herbert 1837). Even under ideal conditions the plants are slow to establish themselves and take a minimum of four years to flower from seed, although five to six years is the more usual time period, but once established, they flower fairly reliably. Experience at Kirstenbosch has been that *C. nobilis* produces two flushes of flowers per year, starting its main flowering period in midwinter and extending to early summer. A second, much smaller flush occurs in autumn.

Forms of this species occurring on primary and secondary coastal dunes should do well in seaside gardens provided they receive some moisture in summer and dappled shade for the hottest part of the day.

A colour surprise occurred recently in *C. nobilis* at Welland Cowley's nursery in the Eastern Cape when a plant with pale yellow flowers and bright green tips arose spontaneously from a batch of seedlings that had been collected in the wild. It is known as *C. nobilis* 'Pearl of the Cape' (Cowley 2006a). The sturdy, leathery leaves of this species prevent attack from lily borer caterpillars and, like those of *C. mirabilis*, the leaves are generally much less susceptible to fungal disease than the other species.

Opposite, above: Orange-pink form of *C. nobilis*

Opposite, below: A colony of *C. nobilis* in fruit

Welland Cowley

Cameron McMaster

85

Ripe berries of *Clivia nobilis*

Clivia robusta

Clivia robusta var. *robusta* from southern KwaZulu-Natal

Clivia robusta B.G.Murray *et al.* var. *robusta* in Botanical Journal of the Linnean Society 146: 369–374 (2004).

Type South Africa, Eastern Cape, 3129 (Port St. Johns): Mount Sullivan, Port St. Johns (–DA), *Truter 4072* (PRU, holotype).

Etymology The specific epithet 'robusta' refers to the robust nature of certain forms of this species.

Distribution Port St Johns in the north-eastern part of the Eastern Cape to Oribi Gorge in the southern part of KwaZulu-Natal, confined to the Pondoland Centre of Endemism.

Habitat Fragmented remnant patches of afromontane forest in grassland, usually restricted to swampy terrain or near water, in deep shade, at altitudes from sea level to 500 m. Encountered where water accumulates in depressions of forest including *Erythrina caffra*, *Phoenix reclinata* and *Syzygium cordatum*.

Flowering period Late March to mid July (mid autumn to midwinter).

Description Extremely variable evergreen, summer-growing, offset-forming perennial 1–2 m high. *Rhizome* 300–400 mm long, with or without buttress roots produced from base of pseudostem. *Leaf sheath* uniformly green. *Leaves* 8–10, strap-shaped, pale to dark green, suberect to arched, weakly canaliculate, 0.4–1.2 m long, 35–90 mm wide, pale white median striation occasionally present in younger leaves, margins flat, apex obtuse-apiculate. *Scape* erect, pale to dark greenish-brown, 0.4–1 m long, 11–17 mm wide, usually much shorter than length of leaves. *Inflorescence* sparse to dense, 10–49-flowered, subtended by 3–4 membranous spathe bracts, 25–30 x 5–8 mm; bracteoles filiform, 11–15 mm long; pedicels sub-erect to erect, pale to dark greenish-brown, 20–40 mm long.

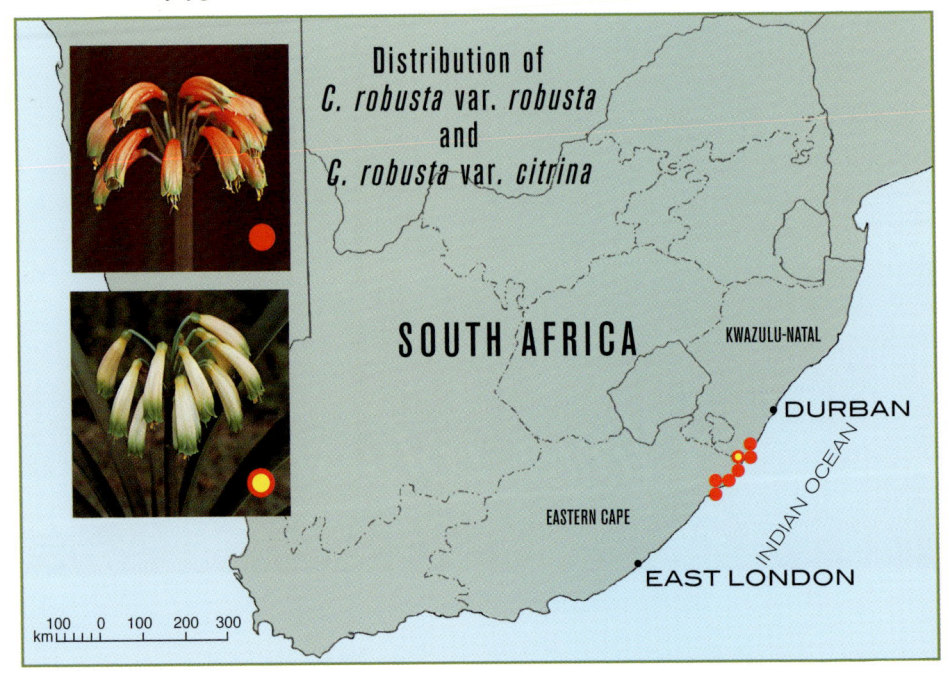

Distribution of
C. robusta var. *robusta*
and
C. robusta var. *citrina*

Clivia robusta var. *robusta* from the Eastern Cape

Clivia robusta var. *robusta* from the Eastern Cape, in the Camphor Avenue at Kirstenbosch

Flowers tubular, curved, drooping, perianth very variable in colour; perianth tube pale to dark orange, orange-red or brick-red, rarely peach pink, 10–14 mm long; outer tepals lanceolate, pale to dark orange, orange-red or brick-red, rarely peach pink, with bright green tips, 21–24 x 6–7 mm; inner tepals narrowly spathulate, protruding slightly beyond outer tepals, with broad, pale to dark orange, orange-red, brick-red or rarely peach pink median keels, flanked by pale greenish-yellow zones, with pale to bright green tips, 24–30 x 14–16 mm. *Stamens* slightly curved, included within perianth or protruding slightly, filaments white or pale green, 29–32 mm long, anthers 4.0 x 1.2 mm. *Ovary* globose, pale to bright green, 4–5 x 5 mm; style slightly curved, white or pale green, protruding slightly beyond perianth, 37–42 mm long, stigmatic branches 1 mm long. *Fruit* a globose or ovoid, dark red or reddish pink pendent berry 15–40 x 10–20 mm, ripening within 9–12 months, from early to midwinter. *Seeds* pale brown, globose or irregularly shaped, 9–16 x 10–18 mm, 1–4 per berry.

Cultivation Resembling a robust form of *C. gardenii*, *C. robusta* var. *robusta* performs admirably in cultivation, both as a garden and container plant. The leaves of the most robust forms, such as those from around Port St Johns, can reach up to 1.2 m in length and are best grown in gardens where they can reach their full potential, whereas smaller forms do well in large pots with a diameter of at least 35 cm. Forms from swampy habitat can be grown in poorly drained soil surrounding shady garden ponds, but they perform equally well under ordinary garden conditions, provided they have sufficient shade and are allowed to remain in the same position for many years.

It is a long-lived plant, flowering reliably every year. A well established clump of an extremely robust form from Port St Johns is to be seen in the Camphor Avenue at Kirstenbosch where it has been growing for over 20 years. The number of flowers per head varies tremendously amongst the numerous local forms of this variety; those from Port St Johns have as few as 10 flowers per head whereas the form from the Bizana district of Pondoland known as *C. robusta* 'Maxima', can have up to 49 flowers (Dixon 2005b).

Clivia robusta B.G.Murray *et al*. var. ***citrina*** Z.H.Swanevelder *et al*. in Bothalia 36 (1): 66–68 (2006).

Type South Africa, KwaZulu-Natal, 3030 (Port Shepstone): Maringo Flats (–CC), *Forbes-Hardinge FH01* (PRU, holotype).

Etymology The specific epithet 'citrina' refers to the pale yellow or lemon-yellow flowers.

Distribution Confined to a single locality in southern KwaZulu-Natal, occurring within a population of *C. robusta* var. *robusta*, in the Pondoland Centre of Endemism.

Habitat Remnant afromontane forest, in swampy terrain, in deep shade.

Flowering period Late May to late June.

Description As for *C. robusta* var. *robusta* (see above), but plants are somewhat shorter (up to 1.6 m high) and flowers are pale yellow or lemon-yellow with pale to dark green tips, not pale to deep orange or orange-red with pale to dark green tips as in var. *robusta*. The berries ripen from early to midwinter, but unlike those of *C. gardenii* var. *citrina*, they are pale to dark yellow, as in *C. miniata* var. *citrina*.

Clivia robusta var. *citrina* from southern KwaZulu-Natal

Cultivation *C. robusta* var. *citrina* is cultivated in exactly the same manner as var. *robusta*. It is rare in cultivation at present and although its natural habitat is swamp-like conditions in overhead shade, it is not a requirement that it be grown under swampy conditions in cultivation.

It performs very well in large containers with a diameter of 35 cm and forms offsets readily. Like those of var. *robusta*, the leaves fall prey to lily borer caterpillars in summer, but appear not to be very susceptible to fungal disease.

Clivia gardenii var. *gardenii* from eastern KwaZulu-Natal

Clivia gardenii W.J.Hooker var. *gardenii* in Curtis's Botanical Magazine series 3, 12: t. 4895 (1856).

Type South Africa, KwaZulu-Natal, precise locality unknown, *Garden s.n.* (K, holotype).

Etymology The specific epithet 'gardenii' commemorates Major Robert Jones Garden of the 45th Sherwood Foresters Regiment.

Common names Major Garden's clivia, Natal drooping clivia, boslelie (Afrikaans), umayime, umgulufu (Zulu).

Distribution Confined to the central, eastern and northern parts of KwaZulu-Natal, extending in an arc from just west of Durban to Howick, Greytown and Empangeni, with disjunct populations around Nongoma, in the Maputaland-Pondoland Centre of Endemism.

Sympatric with *C. miniata* in central, eastern and northern KwaZulu-Natal.

Habitat In deep shade of afromontane forest, on south and south-east facing slopes in shale or sandstone derived soils, in coastal and inland forests and forest patches, amongst boulders and on river banks, from just above sea level to 1200 m. Sometimes epiphytic, growing in forks of trees.

Flowering period Early April to mid-July (mid autumn to midwinter).

Description Highly variable, evergreen, summer-growing, offset-forming perennial 0.8–1.3 m high. *Rhizome* short, subterranean, rarely lithophytic or epiphytic, with thick fleshy roots up to 7 mm wide. *Leaf sheath* uniformly green, or white suffused with dull maroon. *Leaves* 9–17, in a loose fan, strap-shaped, dark

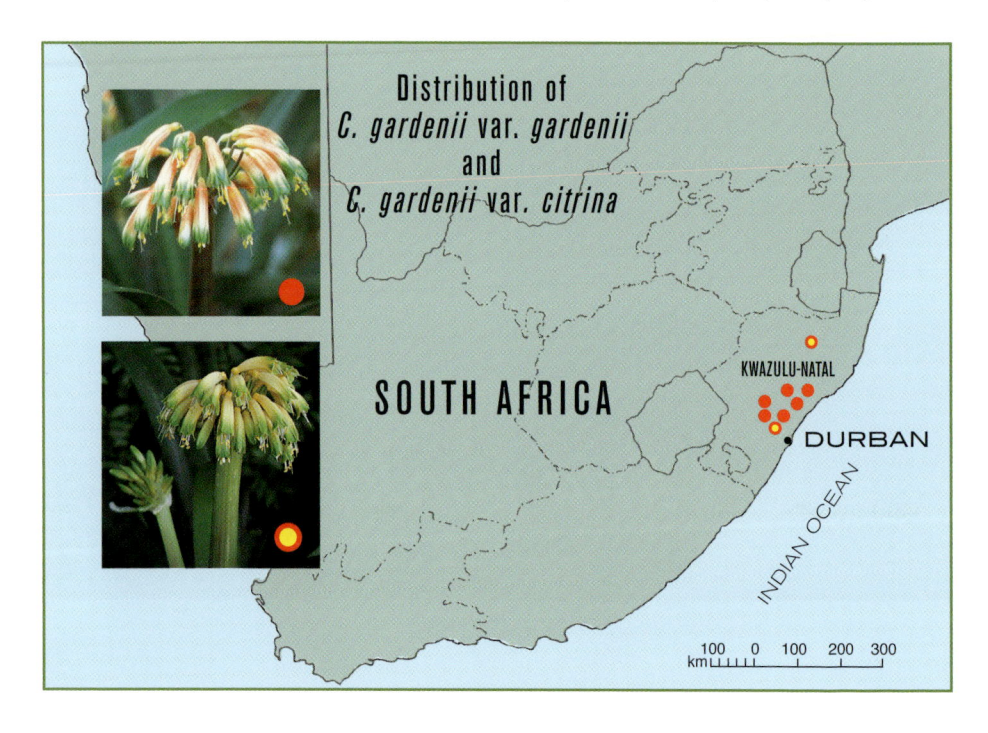

Distribution of
C. gardenii var. *gardenii*
and
C. gardenii var. *citrina*

Clivia gardenii var. *gardenii* from eastern KwaZulu-Natal

green, arching, lax, weakly canaliculate, 650–900 x 30–50 mm, apex acute. *Scape* erect, pale to dark green in deep shade, purplish-maroon in bright light, 250–400 x 15–30 mm. *Inflorescence* sparse to fairly dense, 10–25-flowered, held at or just below the leaves, subtended by 3–4 translucent white, membranous spathe bracts, 50–55 x 7–12 mm; bracteoles linear, membranous, 10–12, translucent white, 28–42 mm long; pedicels sub-erect to erect, pale to dark green or brown, 15–50 mm long. *Flowers* tubular, flaring slightly to markedly towards tips, slightly to distinctly curved, pale to dark orange or orange-red, nodding to pendent; perianth tube 7–10 mm long, pale to dark orange or orange-red; outer tepals lanceolate, canaliculate, pale to dark orange or orange-red with bright green tips, 25–30 x 6–9 mm; inner tepals oblanceolate, pale to dark orange or orange-red, shading to bright green in upper part, slightly to moderately flared, 30–32 x 8–10 mm, protruding slightly beyond outer tepals. *Stamens* slightly to strongly curved, just emerging beyond tip of perianth to well exserted; filaments sturdy, white in lower half, shading to pale green above, 25–40 mm long; anthers oblong, 3–4 x 1–2 mm, pollen yellow at anthesis. *Ovary* globose, pale greenish-yellow to dark green, 4–6 x 4–5 mm, containing 3–4 ovules per locule; style sturdy, slightly curved, white in lower half, shading to green above, extending beyond anthers, 37–49 mm long, stigma tricuspidate, stigmatic branches 2–3 mm long. *Fruit* a globose or ovoid, dark red or reddish-pink drooping berry 20–25 x 18–28 mm, ripening within 9–12 months, in midwinter. *Seeds* pale brown, globose or irregularly shaped, 13–15 x 11–15 mm, 1–4 per berry.

Cultivation *C. gardenii* is an under-utilized garden plant. Cultivated with just as much ease as *C. miniata*, its greatest assets lie in its anomalous flowering period from mid autumn to midwinter, at a time when colour in the garden is in short supply; in the prominent purplish-red scapes of certain forms that provide excellent contrast against its soft, arching green leaves; and in its ability to flower well in even dense shade. A planting of a low-growing form of this variety from Eshowe has been growing in deepest shade in the Dell at Kirstenbosch since 1947. It thrives under a high canopy and is seen to great advantage planted in large drifts under deciduous trees, whose autumn leaves coincide with its striking heads of tubular flowers and often prominently exserted stamens and stigma. It also makes an admirable subject for large containers with a diameter of 35 cm or more, making a striking display on a shady patio. Although offsets are not produced as readily as in *C. miniata*, the plants multiply steadily, and in some instances by stolon formation, once well established. The plants are long lived, flower reliably every year and prefer to remain undisturbed for many years. Their leaves are much less inclined to attack by fungal disease compared with those of *C. miniata*, but are highly susceptible to lily borer caterpillars in summer.

Clivia gardenii W.J.Hooker var. ***citrina*** Z.H.Swanevelder *et al.* in Bothalia 36 (1): 67–68 (2005).

Type South Africa, KwaZulu-Natal, 2731 (Louwsburg): Ngome Forest, (–CD), *Swanevelder* & *Truter ZH10* (PRU, holotype).

Etymology The specific epithet 'citrina' refers to the pale yellow flowers.

Clivia gardenii var. *citrina* from northern KwaZulu-Natal

Distribution Confined to two disjunct populations in eastern and northern KwaZulu-Natal, consisting mainly of yellow-flowered individuals.

Habitat South-facing ravine slopes in afromontane forest.

Flowering period Late May to mid July (late autumn to midwinter).

Description As for *C. gardenii* var. *gardenii*, but flowers are pale to bright yellow or lemon yellow with dark green tips in the wild, or when cultivated in deep shade, not pale to dark orange or orange-red with dark green tips as in var. *gardenii*. However, flower colour in var. *citrina* is unstable and dependent on environmental conditions; under cultivation the yellow colouration is gradually lost with an increase in light, turning to dull orange in bright light (John Winter, personal communication). Unlike those of *C. miniata* var. *citrina* and *C. robusta* var. *citrina*, the berries are not pale to dark yellow, but bright orange-red to dark red (see opposite).

Cultivation *C. gardenii* var. *citrina* is cultivated in exactly the same manner as the var. *gardenii*, and is a magnificent horticultural subject with its dense heads of conspicuous yellow flowers and strongly exserted stamens. Growing to a height of up to 1.3 m, it is more robust than most forms of var. *gardenii*, and flowering from late autumn to midwinter, it provides a welcome splash of colour to the garden at an otherwise dreary time of year. In temperate climates it performs equally well in open ground or large containers, and flowering coincidentally with the ripening bright red fruits of the previous flowering season, it creates a startling colour contrast.

It multiplies rapidly by offset formation. Unlike *C. miniata* var. *citrina*, basal stem pigmentation occurs in *C. gardenii* var. *citrina* plants and seedlings, and is thus not an indication of flower colour.

Clivia gardenii var. *citrina* from northern KwaZulu-Natal

Below: *Clivia miniata* hybrid raised by the author at Kirstenbosch

Opposite: *Clivia miniata* 'Andrew Gibson', an unusual colour form originally collected near Howick by Andrew Gibson

HYBRID, VARIEGATED AND NOVELTY CLIVIAS

HYBRID CLIVIAS

Large numbers of intraspecific hybrids (hybrids between different forms of the same *Clivia* species, mainly *C. miniata*) as well as primary interspecific hybrids (first generation hybrids between different *Clivia* species) have been raised in many parts of the world over the past century and a half, and latterly, the number of advanced hybrids (hybrids between primary hybrids and species) has increased as a result of the rapidly expanding interest in clivias around the world. *Clivia miniata* 'Vico Yellow' is an example of an intraspecific hybrid, *Clivia* Cyrtanthiflora Group is an example of an interspecific hybrid (*C. nobilis* x *C. miniata*), and *Clivia* Minicyrt Group is an example of an advanced hybrid (*C.* [Cyrtanthiflora Group] x *C. miniata*). In addition, the first formally documented natural hybrid, *Clivia* x *nimbicola* has recently been published while a second natural hybrid between *C. gardenii* and *C. miniata* from Stanger in KwaZulu-Natal is illustrated (see page 105).

Clivia x *nimbicola* from the border between northern Swaziland and eastern Mpumalanga

The *Clivia* Cultivar Checklist and Registration process is a means of registering the intraspecific, interspecific and advanced hybrids being grown by providing them with cultivar names in order to record as much information as possible regarding their parentage and history. This information is of great importance in enabling growers worldwide to have detailed information on which to base their growing and breeding decisions. The term 'cultivar' is an abbreviation of 'cultivated variety', and usually refers to a plant raised or selected in cultivation, that retains distinct, uniform characteristics when propagated. All plants carrying a cultivar name must be genetically identical. It can also refer to wild collected forms, such as *C. miniata* var. *miniata* 'Appleblossom'. The process includes, amongst other things, completing a registration application form and providing clear images and a detailed description of the flowering plant. Professional breeders as well as enthusiasts are urged to apply for formal cultivar name registration. In order to qualify for a new cultivar name, it must be unique and distinct from other existing *Clivia* hybrids or wild forms.

Kenneth R. Smith is the International Registrar for the genus *Clivia*. His website is: http://cliviasmith.idx.com.au/

Details of the natural hybrids mentioned above and a selection of artificial hybrids follow.

Natural hybrids

The natural distribution of *C. miniata* is sympatric with that of all four tubular-flowered, summer rainfall *Clivia* species in parts of its range. It is sympatric with *C. nobilis* in the eastern part of the Eastern Cape, with *C. robusta* in the north-eastern part of the Eastern Cape and the southern part of KwaZulu-Natal, with that of *C. gardenii* in a number of localities in central, eastern and northern KwaZulu-Natal, and with that of *C. caulescens* on The Bearded Man Mountain on the border of northern Swaziland and eastern Mpumalanga, South Africa. The first formally documented instance of a naturally occurring *Clivia* hybrid is that between *C. caulescens* and the most northerly populations of *C. miniata* on The Bearded Man Mountain (Rourke 2003a, Swanevelder *et al*. 2006). Theoretically the very different flower shapes of *C. miniata* (trumpet-shaped) and *C. caulescens* (tubular) preclude them from being pollinated by the same pollinator. While butterflies are thought to be the main pollinators of *C. miniata*, and sunbirds the main pollinators of *C. caulescens*, sunbirds and bees do occasionally visit the flowers of *C. miniata*, and bees the flowers of *C. caulescens*. Although the flowering periods of *C. nobilis* and *C. miniata* overlap at a few localities, there are as yet no documented reports of any interspecific hybrids between the two. Natural hybrids are not expected between *C. miniata* and the autumn- and winter-flowering *C. gardenii* and *C. robusta*, as the latter have long finished flowering by the time the *C. miniata* flowering season begins.

Clivia* x *nimbicola Z.H.Swanevelder *et al.* in Bothalia 36 (1): 77–80 (2006).

Type Swaziland, 2531 (Barberton): Bearded Man Mountain (–CB), *Pearton TP01* (PRU, holotype).

Etylomolgy The name 'nimbicola' means 'dweller in the mist', referring to the mist belt habitat where this hybrid and its supposed parents occur. The name *C.* x *nimbicola* refers to all hybrids between

C. miniata (including var. *miniata* and var. *citrina*) and *C. caulescens,* irrespective of which is the pod parent or the pollen parent.

Common name Mist clivia.

Distribution Restricted to a few populations on The Bearded Man Mountain near Baberton on the border between northern Swaziland and eastern Mpumalanga, South Africa, in the Barberton Centre of Endemism.

Habitat Steep rocky slopes and protected forest floor of evergreen afromontane forest, at an altitude of 1445 m, in sympatric populations of *C. miniata* and *C. caulescens.*

Flowering period Mainly May to July, and again from November to December, or sporadically at any time of year.

Description Evergreen, summer-growing, solitary or offset-forming perennial 0.4–1.2 m high. *Rhizome* up to 60 mm wide, extending above ground up to 500 mm high, with thick fleshy roots up to 6 mm wide. *Leaf sheath* uniformly green, dark reddish-green or reddish-maroon. *Leaves* 10–12, strap-shaped, dark green, in a loose fan, arched, surfaces flat, 250–350 x 55–70 mm, apex acute. *Scape* erect to sub-erect, bright green, laterally compressed and distinctly 2–edged, 200–600 x 10–30 mm. *Inflorescence* usually dense, 15–29-flowered, flat-topped, usually held well above leaves, subtended by 2 lanceolate, membranous, translucent white spathe bracts with prominent green margins; bracteoles 4–7, filiform to linear; pedicels spreading to sub-erect, sturdy, dark green, 15–40 mm long. *Flowers* narrowly trumpet-shaped, sub-erect, spreading or drooping; perianth tube funnel-shaped, pale blush pink to blush apricot, 6–10 mm long; outer tepals oblanceolate, pale blush pink to blush apricot, with out without bright green apices, 25–40 x 8–10 mm; inner tepals subspathulate, pale blush pink to blush apricot, with or without bright green apices, slightly to conspicuously flared towards apices, 27–40 x 10–14 mm, protruding slightly beyond outer tepals. *Stamens* straight to slightly curved, included or just emerging beyond tip of perianth, filaments white, 20–30 mm long, anthers oblong, 2 x 1 mm. *Ovary* ovoid, green, 3–4 x 2–3 mm, style slightly curved, white, protruding slightly beyond stamens in late flowering stage, 25–45 mm long, stigmatic branches green. *Fruit* a globose dark red berry 15–20 x 15–20 mm, containing 1–4 seeds. *Seeds* subglobose, creamy-brown, 10–15 x 10–12 mm.

Cultivation *C.* x *nimbicola* is an exceptionally vigorous, easily grown plant with great horticultural potential once material becomes widely available. Morphologically intermediate between *C. caulescens* and *C. miniata* with respect to its leaves, inflorescence and flowers, its attractive flat-topped flower heads bear up to 29 flowers and like *C. caulescens*, it develops a distinct aerial stem with advanced age.

Flowering mainly from late autumn to midwinter, it should bring much-needed colour to temperate gardens of the Southern Hemisphere. It sometimes produces a second flush of blooms in early summer, and sporadic blooms may appear at any time of year. It performs equally well in large containers with a diameter of 30–35 cm. Emanating from relatively high altitude terrain that is subject to occasional snowfalls, the plants may be able to withstand freezing temperatures for short periods.

A natural hybrid between *Clivia miniata* x *C. gardenii* collected by Sean Chubb in Stanger, KwaZulu-Natal in 2003. It was found growing among a population of *C. miniata* plants on a southern slope below a stream, adjacent to a population of *C. gardenii* plants on the northern bank of the stream, flowering in October. An in-depth study of the two populations needs to be undertaken before the hybrid can be formally described.

Above: *Clivia* Cyrtanthiflora Group makes an outstanding garden and container subject

Opposite: Berries of *Clivia* Cyrtanthiflora Group

Artificial interspecific hybrids

Artificial interspecific hybrids and crosses between primary hybrids and other parents are given group names in order for them to be immediately identifiable to growers. Group names currently recognized (Koopowitz 2002), and a new Group name, are:

Group Name	Parentage
Clivia Caulgard Group	*C. gardenii* × *C. caulescens*
Clivia Cyrtanthiflora Group	*C. nobilis* × *C. miniata*
Clivia Minicyrt Group	*C.* [Cyrtanthiflora Group] × *C. miniata*
Clivia Minigard Group	*C. gardenii* × *C. miniata*
Clivia Minilescent Group	*C. miniata* × *C. caulescens*
Clivia Nobilescent Group	*C. nobilis* × *C. caulescens*
Clivia Noble Guard Group	*C. gardenii* × *C. nobilis*
New Group Name	**Parentage**
Clivia Minirabilescent Group	*C. mirabilis* × *C. miniata*

***Clivia* Cyrtanthiflora Group**

Type Figure in *Flore des Serres* (Series II) 87, plate 1877 (1869).

Synonyms *Imatophyllum cyrtanthiflorum* van Houtte, Flore des Serres (Series II) 87: t. 1877 (1869); *Clivia* x *cyrtanthiflora* (van Houtte) H.P.Traub in Plant Life 32: 57 (1976).

Etymology The name 'cyrtanthiflora' refers to the curved, pendulous flowers, reminiscent of those of many *Cyrtanthus* species. 'Cyrtanthiflora Group' refers to all primary hybrids between *C. miniata* and *C. nobilis*, irrespective of which the pod parent or the pollen parent is.

Flowering period Late June to late July (midwinter), and early November to mid December (early summer), or sporadically at any other time of year.

Description Evergreen, summer-growing, strongly offset-forming perennial 0.5–1 m high with a short subterranean rhizome. *Leaf sheath* uniformly green. *Leaves* 10–20, strap-shaped, somewhat leathery, dark green, sub-erect to arched, weakly canaliculate, 0.5–1 m long, 30–55 mm wide, margins minutely serrated, apex usually sub-acute, rarely obtuse or weakly notched. *Scape* erect to sub-erect, pale to dark green, laterally compressed and distinctly 2-edged, narrowing markedly towards apex, 30–50 x 20–45 mm. *Inflorescence* fairly dense, 15–40-flowered, usually held well below leaf tips, subtended by 3–4 lanceolate to oblong, translucent white spathe bracts, 25–45 x 10–20 mm, membranous becoming papery with age; bracteoles filiform to linear, translucent white, 25–35 mm long; pedicels weakly arched to erect, dark green, 25–45 mm long. *Flowers* weakly tubular to narrowly funnel-shaped, slightly curved, drooping; perianth tube funnel-shaped, pale to dark orange-red, 5–10 mm long; outer tepals lanceolate, pale to dark orange-red, with pale green tips fading to orange-red with age, 35–40 x 8–10 mm; inner tepals spathulate, pale to dark orange-red, with pale green tips fading to orange-red with age, 38–42 x 15–20 mm, protruding slightly beyond outer tepals. *Stamens* slightly curved, included within perianth, filaments yellowish-white, 35–40 mm long; anthers oblong, 2–3 x 1 mm. *Ovary* globose, dark green, 4–5 x 3–4 mm; style slightly curved, protruding well beyond perianth, yellowish-white with reddish uppermost portion, 52–58 mm long, stigmatic branches very short, yellowish-white. *Fruit* a globose, bright red berry 15–20 x 15–25 mm. *Seeds* globose, pale brownish-white, 5–7 x 5–7 mm, 1–3 per berry.

Often misidentified as *C. nobilis*, plants of *C.* Cyrtanthiflora Group are easily distinguished by their larger overall size, having much longer, wider flowers that are weakly flared towards their tips with a well exserted style, borne on much longer pedicels, and by their less leathery, broader leaves that usually have sub-acute apices. Many hybrids of the *C.* Cyrtanthiflora Group are noted for the attractive contrasting colour of the inner and outer tepal surfaces.

Cultivation This vigorous, floriferous, long-lived hybrid performs extremely well in cultivation. Most closely resembling *C. nobilis*, its inflorescence is certainly more desirable than the latter species, and some of the better forms are almost as showy as *C. miniata*. The larger forms are suited to mass planting in garden beds and all forms make outstanding container plants.

Although only one inflorescence is usually produced per plant at a time, the flowers are extremely long lasting and if not pollinated, remain attractive for 18 days or more. In addition to the main flowering period in midwinter, it usually produces a second flush of blooms in early summer. Its dark red, shiny fruits are an added attraction and it readily produces offsets.

Especially vigorous progeny is obtained from crosses where *C. miniata* is used as the pod parent, sometimes flowering in as little as three and a half years. Variegated forms have also been raised.

Clivia **[Minigard Group]** 'Journey'

Bred by Yoshikazu Nakamura in Japan, the parentage of this second-generation, variegated hybrid is (*C. miniata* x *C. gardenii*) x self, and was named 'Journey' by Shigetaka Sasaki.

Clivia **[Minilescent Group]** 'Mandala'

A spectacular *C. miniata* x *C. caulescens* hybrid bred in Japan by Yoshikazu Nakamura, and named 'Mandala' by Shigetaka Sasaki. This vigorous plant has broad leaves, a stout, broad scape and a well rounded inflorecence. The flowers are loosely arranged in the circular manner of a mandala, with the pinkish-orange upper surface of the tepals providing excellent contrast against the yellowish-orange lower surface. It flowers in late winter and early spring, and again in midsummer.

Left: *Clivia* [Minigard Group] 'Journey'

Below: Yoshikazu Nakamura admires his *Clivia* [Minilescent Group] 'Mandala'

Above: *Clivia* [Minilescent Group] 'Mandala' bred by Yoshikazu Nakamura in Japan
Opposite: *Clivia* Minirabilescent Group (see also page ii)

Clivia Minirabilescent Group

Bred by John Winter, two plants of the first cross between *C. miniata* (pod parent) and *C. mirabilis* (pollen parent) flowered at Kirstenbosch for the first time in October 2006. The fairly dense inflorescence is held just above the leaves, atop an erect, dark green or pale maroonish-brown scape. There is considerable variation in pedicel length and orientation.

On opening, the spreading to pendent flowers are pale salmon pink with pale yellow tips, maturing to a rich pinkish-red. The seedlings are fast growing and flower in their third season.

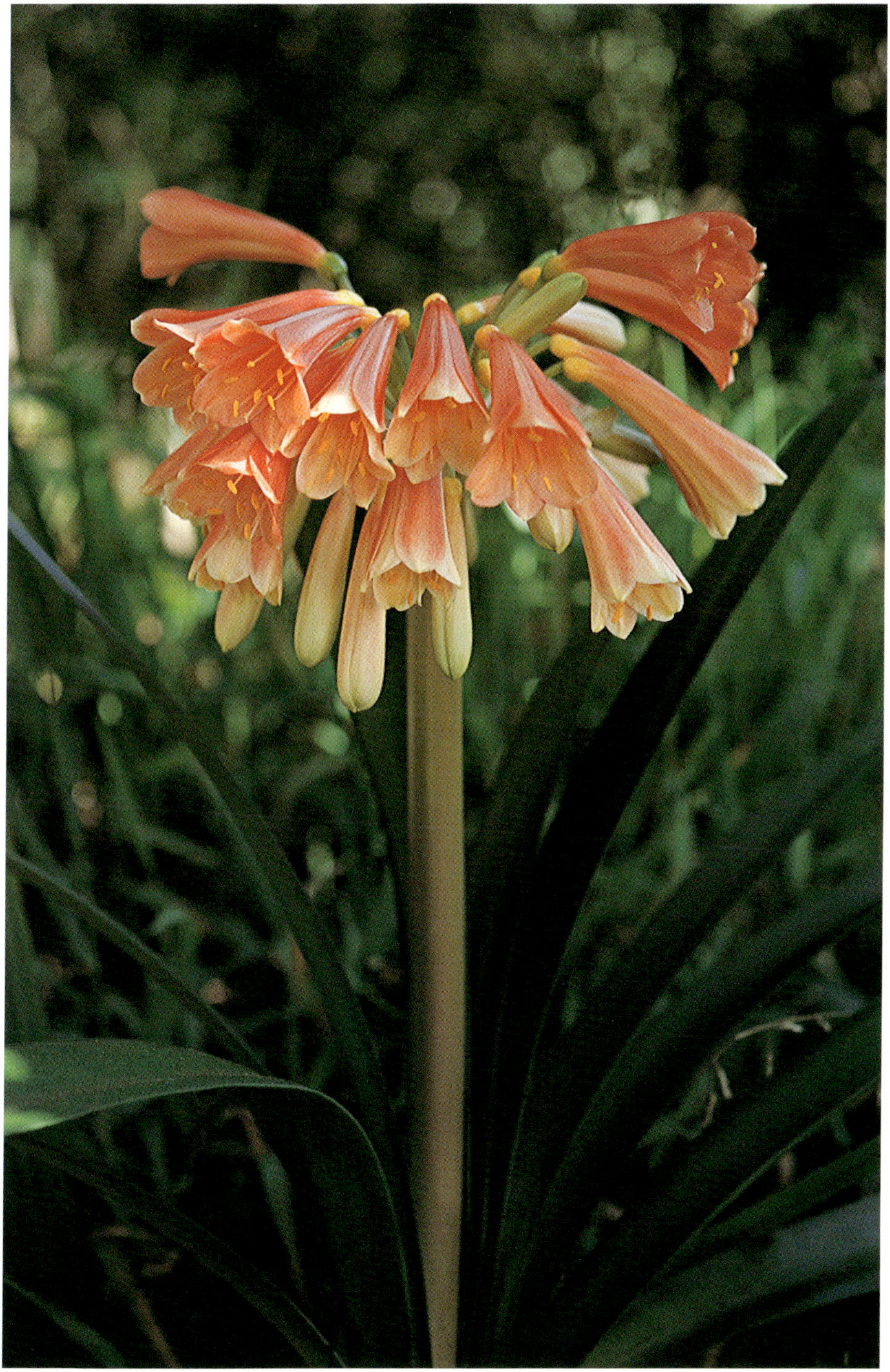

***Clivia* [Group unknown]** 'Crayon'

Bred by Yoshikazu Nakamura and named 'Crayon' by Shigetaka Sasaki, this second-generation hybrid of unknown parentage has very narrow foliage and very small flowers.

***Clivia* [Group unknown]** 'Soko Jiro'

Bred by Yoshikazu Nakamura in Japan, this is a second-generation hybrid of unknown parentage.

Artificial intraspecific hybrids

Clivia miniata Cameron Peach strain

Raised by Cameron Onie and Michael Christie at Tipperary Nursery above Nelspruit in Mpumalanga in the early 1980s, the first plant appeared from a batch of seed-grown Belgian Hybrids that had been open-pollinated with various other forms of *C. miniata*. The plant was isolated and propagated from seed by 'selfing', a large percentage of which were true to type. Slight variation in flower form and flower colour exists in the Cameron Peach strain, a selection of which has slightly darker peachy-apricot flowers and was named 'Cheryl Apricot' (Christie 2006). Cameron Peach strain is a floriferous plant with very showy, upward facing flowers and broad, dark green foliage. With the exception of the specially selected 'Cheryl Apricot', remaining individuals from this strain are best known as Cameron Peach strain, as no particular individual was selected and given the cultivar name 'Cameron Peach'.

Opposite above: *Clivia miniata* 'Cheryl Apricot'

Opposite below: *Clivia* [Group unknown] 'Crayon'

Right: *Clivia* [Group unknown] 'Soko Jiro'

Below: *Clivia miniata* Cameron Peach strain

Shigetaka Sasaki

Clivia miniata Daruma strain

The Japanese term 'Daruma' is a general one used for many *C. miniata* dwarf hybrids raised in Japan. They typically have broad, short leaves with rounded apices that point downwards and pale to dark orange blooms. This unusual and beautiful yellow-flowered plant was bred by Toshio Koike in Japan.

Clivia miniata 'Hirao'

Bred by Toshio Koike in Japan, this beautiful green hybrid was bred from Group 2 yellows. After breeding a number of dark orange *C. miniata* hybrids with strong green centres, and using these as pod parents, he crossed them with a Group 2 yellow (pollen parent). He then crossed the F1 siblings, and from this cross he obtained the clear green flowers (Sasaki 2005). According to Koike, the flowers open green and remain so until anthesis, after which the green gradually fades to cream. He says the flowers revert to cream when their nutritional requirements are neglected, adding that the plant requires regular application of nitrogenous fertilizer. The cultivar name 'Hirao' honours the late Dr Shuichi Hirao, a very knowledgeable Japanese plantsman.

Right: *Clivia miniata* Daruma strain (yellow form)

Below: *Clivia miniata* 'Hirao'

Clivia miniata 'Kirstenbosch Supreme' bred by the author at Kirstenbosch (see page 116)

Clivia miniata Kirstenbosch Supreme strain

Bred by the author at Kirstenbosch in the late 1980s, 'Kirstenbosch Supreme' is a selection from a cross between *C. miniata* var. *citrina* 'Kirstenbosch Yellow' (pod parent) and *C. miniata* var. *citrina* 'Natal Yellow' (pollen parent) (Duncan 1985, 1992). 'Kirstenbosch Yellow' is a Group 1 yellow and 'Natal Yellow' is a Group 2 yellow, and crosses between these two groups always produce orange-flowered progeny. The most attractive plant from the cross was selected and named 'Kirstenbosch Supreme' by John Winter (2000) (see also title page and page 115). 'Kirstenbosch Supreme' is a magnificent, large-flowered plant with well re-curved tepals and a large, nicely rounded inflorescence borne on a sturdy scape. The flowers are azalea-scented and two flower heads per plant are often produced. It is not a repeat flowerer but multiplies fairly well by offset formation. Remaining individuals from this strain are best known as Kirstenbosch Supreme strain.

Clivia miniata Pat's Gold strain

Another hybrid raised by the author at Kirstenbosch in the late 1980s, 'Pat's Gold' is a selection from a cross between *C. miniata* var. *citrina* 'Butter Yellow' (pod parent) and *C. miniata* var. *citrina* 'Natal Yellow' (pollen parent). As both parents are Group 2 yellows, the progeny was all yellow. A number of plants of this cross were made available at the annual Plant Sale of the Botanical Society of South Africa, one of which was purchased by Pat Gore of Pretoria, which became known as 'Pat's Gold'. Another individual from this cross was illustrated on the front cover of *Clivia* 2 (2000). One of the best of the Group 2 yellows, it is a vigorous plant with a full rounded head of dark golden yellow blooms, often producing two inflorescences per season.

Clivia miniata 'TK Best Bronze'

A fine bronze hybrid bred in Japan by Toshio Koike and named by Shigetaka Sasaki.

Opposite: *Clivia miniata* Pat's Gold strain

Below: *Clivia miniata* 'TK Best Bronze'

Above: *Clivia miniata* 'Vico Yellow'

Opposite: *Clivia miniata* 'TK Yellow'

Clivia miniata 'TK Yellow'

An outstanding hybrid bred in Japan by Toshio Koike, a cross between his 'original yellow' and a greenish-yellow clone from Dr Hirao, both of which are Group 2 yellows. The tepals have prominent green median keels and the flowers open a clear bright green, maturing to rich yellow.

Clivia miniata 'Vico Yellow'

A well known hybrid with a long and complicated history, inadvertently raised by Sir Peter Smithers (the multi-talented gardener, hybridist, barrister, historian, naval officer, diplomat, Member of Parliament, cabinet minister, photographer and author) in the late 1970s at his home 'Vico Morcote' in Switzerland. In November 1970, Sir Peter obtained three *Clivia miniata* hybrid clones named 'Kewensis A', 'Kewensis B' and 'Kewensis Cream' that had originally been bred by Charles Raffill at Kew who had been attempting to recover a 'lost' yellow-flowered clone from Eshowe by selecting back. The 'Kewensis Cream' clone produced cream-coloured flowers and yellow fruits, and Sir Peter used its pollen on the orange-flowered 'Kewensis A' and 'Kewensis B', the seedlings of which all produced orange-flowered plants with the exception of a single plant that, together with an excessive number of other seedlings that resulted from these crosses, was relegated to a lowly position under a greenhouse bench. When it flowered for the first time in 1978, surprisingly it had shapely creamy yellow flowers with deeper yellow throats. In time the plant produced offsets, one of which was sent to the highly respected Japanese plantsman Dr Shuichi Hirao. Following Dr Hirao's premature death, the plant was obtained by Yoshikazu Nakamura, by which time it had become known as 'Smithers Yellow' in Japan.

In granting Nakamura permission to make use of the plant in his breeding programmes, Sir Peter requested that it be known as 'Vico Yellow', the name by which it is now recognized (Smithers 1995, 2000). Its excellent genetic attributes include the production of large flowers with good colour, tepals that are well re-curved, and pollen that is highly fertile, resulting in a good seed-set. The plant was registered in Japan by Miyoshi & Co. who propagated it by tissue culture, and became the first to successfully propagate any member of this genus by this method. It has since been widely distributed. Subsequently, another Smithers seedling produced yellow flowers of a darker hue, known as 'Vico Gold', that is not yet widely grown.

Clivia miniata 'Vico Peach'

A superb hybrid between *C. miniata* 'Vico Yellow' (pod parent) and *C. miniata* 'Chubb's Peach' (pollen parent), bred by Charl Malan in South Africa.

Clivia miniata 'White Ghost'

Yoshikazu Nakamura is the breeder of this striking cross between *C. miniata* 'Vico Yellow' and a hybrid between an orange *C. miniata* and *C. miniata* var. *citrina*.

Right: *Clivia miniata* 'White Ghost'
Below: *Clivia miniata* 'Vico Peach'

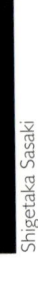

A selection of clivias from Sean Chubb's collection (continued on page 122)

Above: *Clivia miniata* 'Vico Yellow'

Right: *Clivia miniata* 'So Excited'.
A plant from Sean Chubb's breeding programme of group 1 yellows, it has *C. miniata* var. *citrina* 'Aurea', the well known Australian yellow as one of its parents

Right, below: *Clivia miniata* 'Skwebizi Bicolor'.
This outstanding apricot form of *C. miniata* was originally collected by Kobus Steenkamp as a seedling in the Skwebizi Valley near Vryheid

Photographs by Sean Chubb

Above: *Clivia miniata* 'Soft Touch'.
A seedling bred by Sean Chubb from the breeding line *C. miniata* var. *citrina* 'Natal Yellow' x *C. miniata* 'Naude's Peach'

Left: *Clivia miniata* 'Determination'.
A seedling bred by Sean Chubb from selfing the apricot *C. miniata* 'Morné Bouquex'

Left: *Clivia miniata* 'Mars'.
A select seedling from a line bred group originating from seed obtained from Yoshikazu Nakamura. Bred by Sean Chubb

Photographs by Sean Chubb

Opposite: *Clivia miniata* 'Fukurin-fu' (centre) and *C. miniata* 'Akebono-fu' (top left corner) at Yoshikazu Nakamura's Clivia Breeding Plantation, Japan

VARIEGATED CLIVIAS

When different colours are displayed in the leaves of plants, they are referred to as being variegated, and the variegation arises for different reasons. Most variegated clivias have regular variegations, recognized as parallel lines of white or yellow, or a combination of these. True variegation in clivias occurs as a result of a mutation in the DNA of chloroplasts (the organelles in cells that conduct photosynthesis), resulting in the absence of the green pigment chlorophyll in the mutated chloroplasts, which in turn results in cells with mixed chloroplasts (Kirk & Tilney-Basset 1978). This absence may be total or partial, in which case the variegation varies from white to greenish yellow, respectively. Variegated clivias are generally less vigorous than normal green plants because there is less energy-absorbing, food-producing (i.e. green) tissue for the same mass of plant. Variegated plants are most commonly encountered in *Clivia miniata* in cultivation, but are also known from the wild, as well as in *C. caulescens*. For information on cultivation and propagation of variegated clivias, see page 135.

Zonneveld (2006) identified five main origins of leaf variegation. In **natural variegation**, all plants of a particular species exhibit the same variegation and can be grown 100% true to type from seed, such as *Dracaena surculosa* that has yellow spots on its leaves.

Environmental variegation is not true variegation, and may arise as a result of trace element deficiency, incorrect application of pesticides, toxic compounds in the soil, insect damage or viral infection. Leaves infected with a virus can be recognized as uneven blotches or streaking in the form of broken longitudinal lines that often extend sideways into the leaf. Contrary to popular belief, variegation in the leaves of clivias is not usually caused by viral infection. When variegated leaves arise as a result of viral infection, the variegated portion of the leaf often develops dead tissue, and the condition is infective if the sap comes into contact with that of unvariegated leaves.

Temperature or light-dependent variegation is a special type of environmental variegation, in which the lighter leaf colour only occurs at a particular time of year, or when the leaf has reached a certain age. An example of the latter is 'Akebono' variegation in clivias, in which emerging leaves are devoid of yellowish-white bands and only develop this character later on.

Yellow leaf variegation is also not true variegation, but occurs as a result of a mutation that is dominant for the colour yellow, but recessive for its lethal character. Examples of such plants include perennials like *Hosta* and *Sedum*.

True variegation occurs when the leaves contain genetically different cell layers, resulting in genetic mosaics that can arise when a cell in the apical meristem undergoes a spontaneous mutation, i.e. a change in DNA occurs. The mutation produces a cell layer that is partly or completely genetically different, resulting in a 'streaked' plant. Truly variegated plants are almost always unstable and have to be propagated vegetatively.

Chinese and Japanese growers are the leaders in the field of variegated clivias, and in general, they regard dwarf forms more highly than medium-sized or large forms. The rather different directions of development in these two countries have resulted in classification of different groups of variegation. In China they are classified according to three main categories of short, medium and long leaves, with those having the shortest, widest leaves arranged as symmetrically as possible being most highly prized. In Japan, emphasis is placed not on leaf size, but on the distinctive pattern of arrangements that they have developed.

So desirable have variegated clivias become in China that they were the subject of an entire book (Shiang & Song 1999). One of the most popular Chinese variegations is 'Light of Buddha' (see below).

Clivia miniata 'Light of Buddha'

A very striking, broad-leafed mutation developed fairly recently in China, the almost luminous yellowish-green colouration only appears in young leaves after three to six normally pigmented leaves have formed. The variegation occurs in the lower, central and above-central part of the leaf, while the uppermost part is normally pigmented, and is light dependent, requiring high light intensity for it to fully manifest. It then fades to normal pigmentation as the leaves age. Berry colour varies in shades of pale yellow to pinkish-red.

Opposite above: *Clivia miniata* 'Akebono-fu'
Opposite below: *Clivia miniata* 'Light of Buddha'

Seven major variegation types are recognized in Japan (Sasaki 2001, 2003):

Clivia miniata 'Akebono-fu'

At first, the new leaves are normally pigmented, but as they grow older they develop yellow or yellowish-white horizontal bands of varying width across the leaf, in one or more areas of the leaf. 'Akebono', meaning 'sunrise' in Japanese, refers to these bands. In certain forms, the variegation is only visible in young leaves and fades with age, but in others it is permanent, as a result of selective breeding. However, plants with stable variegation patterns restricted to a particular area of the leaf have not yet been isolated. 'Akebono-fu' is an inheritable characteristic but the plants are best propagated vegetatively as success from seed is limited. It is highly sensitive to sunshine or excessive light, and is best grown under relatively low light conditions.

Clivia miniata 'Fukurin-fu'

A most striking plant, the leaves have broad white or yellowish-white bands around the margins of the entire leaf. It is best propagated vegetatively as the berries are usually white, resulting in albino seedlings.

Right and below: *Clivia miniata* 'Fukurin-fu' at Yoshikazu Nakamura's Clivia Breeding Plantation, Japan

Clivia miniata 'Genpei-fu'

The leaves are more or less evenly divided into two parallel, longitudinal zones, one being solid white, the other solid green or a mixture of green and white stripes. When the young leaves emerge they are half greenish-yellow and half dark green, the greenish-yellow portion turning white with age. Plants of this clone command high prices and are not yet readily available.

Clivia miniata 'Negishi-fu'

The plants originate from crosses between different forms of *C. miniata* 'Shima-fu'. The new leaves display extremely narrow, pale to dark green continuous lines against a bright yellowish-green background. In some forms the variegation is lost in older leaves, which eventually turn dark green, while in others it is a permanent feature.

Above: *Clivia miniata* 'Genpei-fu'
Above right: *Clivia miniata* 'Negishi-fu'
Below: *Clivia miniata* 'Naka-fu'

Clivia miniata 'Naka-fu'

The leaf patterning consists of very narrow, parallel, continuous or broken dark green lines in the centre of the leaf on a yellowish-green background.

The pedicels and berries in all forms of the plant are yellowish-green or rarely green, and as the berries mature, they change to purplish-red. Variegation is rarely apparent in the leaves of young seedlings, but as they mature, the yellowish-green colour appears. It is an inheritable characteristic.

Clivia miniata 'Shima-fu'

This is the most commonly grown type, frequently seen for sale in Japanese garden centres. It has relatively broad leaves with rounded apices, downward facing leaves and an irregular pattern of continuous, usually narrow, pale yellow, white, and pale to dark green parallel stripes. Wide variation occurs in the colour of 'Shima-fu' berries, from completely white or yellow to heavily green-striped. Seeds harvested from yellow and white berries will be albinos and will ultimately die, whereas those harvested from well-striped berries will usually yield well-variegated leaves. It is an inheritable characteristic (see next page).

Above and below: *Clivia miniata* 'Shima-fu'
Opposite above: *Clivia miniata* 'Tora-fu' ('Taihou')
Opposite centre: *Clivia miniata* 'Tora-fu'
Opposite below: *Clivia miniata* 'Ito fukurin-fu'

Clivia miniata 'Tora-fu' ('Taihou') (right)
Named 'Tora-fu' in reference to the markings of a tiger, the variegation takes the form of irregular longitudinal stripes and blotches of yellow that vary greatly from one leaf to another. This environmentally induced variegation is caused by viral infection, hence the additional name 'Taihou'.

Two new Japanese variegation types have recently appeared (Shigetaka Sasaki, personal communication).

Clivia miniata 'Ito fukurin-fu' (below)
In new leaves, the variegation appears as a continuous bright greenish-yellow margin, but is lost with age.

Clivia miniata 'Tora-fu' (right)
A much more attractive variegation than the virus-induced 'Tora-fu' (Taihou) pattern, the leaf has continuous or broken horizontal, greenish-yellow bands against a dark green background.

NOVELTY CLIVIAS

Novelty clivias are usually regarded as selected mutations or hybrids of cultivated origin that exhibit strange or unusual new features in their leaves and/or flowers. In the initial stages, they typically appeal to a very small minority of enthusiasts and an even smaller number of commercial growers, and are thus expensive and difficult to obtain. However, refinement of desirable characters through careful breeding and selection can result in highly ornamental lines and the creation of profitable avenues in clivia breeding, as has occurred in China and, to a lesser extent, Japan, with the development of *C. miniata* dwarf mutations and hybrids with short, very wide leaves, for example, in which the inflorescence is of little or no importance. Amongst these dwarfs, Chinese breeders place great emphasis on the production of plants with a 'layered' effect, with the leaves arranged strictly above one another, in two well defined rows, and on vein colouration that contrasts sharply against the basic leaf colour. Japanese breeding concentrates less on the arrangement of the leaves and more on the production of plants with uniformly dark green, shiny leaf colour.

Below: A *Clivia miniata* multitepal of the chrysanthemum type, at the Clivia Breeding Plantation, Japan

Shigetaka Sasaki

The Japanese have also produced a number of ultra-dwarf *C. miniata* hybrids with sharply pointed leaf apices, known as 'pygmies' among breeders in the west. Current well-known flower novelties include those with double the normal number of tepals or those with numerous extra tepals ('multitepals'), as well as those with unusually narrow, spidery ('frats') tepals. The fascination with a particular novelty clivia, like those of other ornamental plants, is subject to fashion, and usually lasts as long as it remains difficult to obtain. Once widely available, they of course cease to be novelties.

Shigetaka Sasaki

Above: A *Clivia miniata* 'Frats' tepal, and below, *Clivia miniata* 'pygmies' at the Clivia Breeding Plantation, Japan

Below: *Clivia miniata* thrives in dappled shade and can be very successfully inter-planted with other shade-loving plants like *Crassula multicava*, seen here sprawling over a dry stone wall

Opposite: *Clivia* Cyrtanthiflora Group

CULTIVATION

Grown for their strong, warm colours and decorative, strap-shaped, evergreen leaves, clivias are among the most easily cultivated bulbous plants, their greatest asset being their willingness to produce attractive inflorescences in shade. Clivias are 'tough' plants that can be handled almost like succulents, and are noted for their tolerance of a dry atmosphere and drought during their semi-dormant phase. They continue to gain recognition throughout the world as one of the most desirable evergreen, geophytic plants and in countries that experience prolonged frost they make outstanding indoor and conservatory plants. Depending on climatic conditions, they are ideal garden or container plants, and *C. miniata* is also a useful cut flower. The tubular-flowered species attract sunbirds to the garden and the eye-catching ripe berries provide many months of additional colour.

Aspect and climate

Dappled shade or bright, filtered or indirect light is the most suitable location for growing clivias outdoors. They will also thrive in heavy shade, but under these conditions most species will flower erratically, if at all. An exception is *C. gardenii* that will often flower in even heavily shaded conditions, although its flowers tend to be somewhat paler than normal.

In the Southern Hemisphere a south- or south-east-facing aspect is best, whereas a north- to north-east-facing aspect is recommended for the Northern Hemisphere. When grown outdoors, all the species can take a couple of hours of direct early morning sun, but should have shade for the rest of the day. Excessive exposure to sun results in rapid and severe scorching of the leaves and flowers, as does prolonged frost. The plants like a protected environment, free from strong wind. Clivias are excellent subjects for large containers on a shady patio, and all can be grown very successfully indoors in a position receiving good light but not direct sunlight. Although often encountered growing under fairly humid conditions in their natural habitat, they like cool, moist conditions and seem to thrive in areas with dry summers and cool, wet winters, such as in the southern suburbs of the Cape Peninsula. Summer-growing clivias grown in extremely humid climates will usually produce reasonable vegetative growth, but flower poorly, while *C. mirabilis*, the only winter-growing clivia, is intolerant of high humidity. Clivias are also intolerant of very hot temperatures as are experienced in the interior of South Africa.

Below left: *Clivia mirabilis* requires excellent drainage and is intolerant of high humidity

Below right: Variegated clivias require lower light intensity for their leaves to develop to their full potential and are best grown in positions where they will receive bright, filtered light in the morning and afternoon shade

Sean Chubb

An ideal aspect for all species is in the dappled shade of evergreen or deciduous trees with a high canopy, or alternatively under shade-cloth with an 80% light exclusion factor, which is ideal for both seedlings and mature plants. *C. mirabilis* and coastal forms of *C. nobilis* are more sun-tolerant and able to withstand higher light intensity. An ideal outdoor temperature range for clivias is 10–25 °C, but they can quite easily tolerate 5–35 °C.

Variegated clivias can be grown in similar conditions to those of ordinary clivias, but there are some differences in temperature and light requirements. They require slightly lower light intensity than ordinary clivias do in order for their leaves to develop to their full potential and are therefore ideal as house plants. They are best grown in positions where they will receive bright, filtered light in the morning and afternoon shade. Plants with short, broad leaves are ideal for windowsills while larger forms can be displayed on patios, preferably facing south-east in the Southern Hemisphere. In the Northern Hemisphere, they should face north-west. Their leaves are much more sensitive to sunburn than ordinary clivias and those with broad bands of white or yellow are much more susceptible to sunburn than those with thin bands of alternating green or yellow. The plants should not be placed in positions where containers may heat up too quickly. An ideal temperature range for variegated plants is 12–25 °C, while temperatures above 30 °C will result in poor growth performance. Like ordinary clivias, variegated plants do not thrive under conditions of high humidity, and prefer a slightly lower humidity range (60-70%) than ordinary plants. Without a certain length of exposure to bright filtered light, the yellow or white stripes in the leaves will begin to deteriorate. However, if they are exposed to direct sun for too long, these areas of the leaf burn quickly.

Hardiness

All six *Clivia* species are frost tender and are best grown in large tubs or pots in a conservatory, cool or warm greenhouse, or inside the home in very cold climates. Clivias can take temperatures down to zero degrees for periods of one to two days, but are severely affected by prolonged exposure to temperatures below zero. In frost susceptible areas, plants can be covered with frost cloth or a frost blanket, known in the trade as 'Crop Cover', that is best double-folded in order to prevent damage under severe conditions. In marginal areas they can be grown under overhanging eaves that provide some protection against frost and sunburn. Although most populations of *C. caulescens* and *C. miniata* occur in frost-free areas, plants occurring at an altitude of 1455 m and experiencing occasional snowfalls on The Bearded Man Mountain on the border between South Africa and northern Swaziland, should have a greater cold tolerance than those found in coastal forest of the Eastern Cape and KwaZulu-Natal.

Clivias in the garden

With the exception of *C. mirabilis,* which is best suited to cultivation in containers, *Clivia* species are suitable for outdoor cultivation in temperate climates; and it is the large-flowered, eye-catching *C. miniata* that lends itself most admirably to general garden use. It is seen to best advantage planted in large drifts under high canopied deciduous or evergreen trees and is also displayed to great advantage in well placed groups, interplanted with other indigenous, shade-loving evergreen or summer-growing geophytes like *Crinum moorei, Crocosmia aurea, Drimiopsis maculata, Scadoxus membranaceus, S. multiflorus* subsp. *katharinae* and *Veltheimia bracteata.* Suitable companion plants amongst the herbaceous perennials include *Asparagus densiflorus, Crassula multicava, Impatiens flanaganiae, I. sylvicola, Knowltonia vesicatoria, Laportea grossa* (stinging nettle), *Peperomia retusa,* numerous low-growing *Plectranthus* species, especially *P. ambiguus, P. ciliatus, P. fruticosus, P. oertendahlii* and *P. verticillatus,* as well as *Streptocarpus formosus, S. primulifolius* and *S. rexii.* It is also most attractively displayed in association with the statuesque and shade-loving *Dracaena aletriformis,* shade-loving cycads like *Encephalartos villosus, E. paucidentatus* and *Stangeria eriopus,* or underneath shady tree ferns such as *Cyathea capensis* and *C. dregei,* or in association with medium-sized ferns like *Blechnum tabulare, Rumohra adiantiformis* and *Todea barbara,* and low-growing species including *Adiantum aethiopicum* and *A. capillus-veneris.*

C. miniata is so versatile that it is recommended for any difficult, shady part of the garden and will grow in almost any well drained soil. In a garden setting, the pale yellow flowers of *C. miniata* var. *citrina* often look somewhat insipid, but contrast wonderfully well with orange and red forms that flower at the same time. The less showy, pendulous-flowered *C. caulescens, C. gardenii* and *C. nobilis* are better suited to well drained, shady rock garden pockets than to mass displays. Forms of *C. robusta* from swampy terrain can be grown in poorly drained parts of the garden like the edges of a shady garden pond.

Clivias can be grown in heavy shade where their decorative leaves remain attractive throughout the year, but will then not flower as well as those placed in dappled shade; with the exception of certain forms of *C. gardenii* (such as from the forests of Eshowe) that are so obliging that they flower in even the heaviest shade, although flower colour tends to be rather washed out.

With a carefully assembled collection of the many different forms of each species, it is possible to have a continuous and overlapping display of clivias in flower in the garden for eight months, from mid autumn (late March) to early summer (mid November). First to come into flower from late March to early August is *C. robusta*. From early April to mid July *C. gardenii* is in flower, followed in mid July to early December by *C. nobilis*, then from mid August to mid November by *C. miniata*, and from early September to early November by *C. caulescens*. In addition, *C. nobilis* and *C. caulescens* frequently produce a second flush of blooms in autumn, as well as sporadic blooms throughout the year, and the brightly coloured ripe fruits of all the species provide long-lasting colour for many months, often contrasting beautifully with flower heads of the current season.

Opposite: *Clivia miniata* will grow in almost any well drained soil

Below left: *Clivia gardenii* flowers from autumn to midwinter

Below right: *Clivia caulescens* is best suited to shady rock garden pockets and frequently produces a second flush of blooms in autumn

As clivias greatly resent root disturbance, their position in the garden should be seen as long-term in order for them to develop to their full potential. They can be left in the same position for up to ten years or more, until they become too overcrowded and flowering performance diminishes. Best flowering results are always obtained from well-established clumps. They prefer slightly sloping ground for improved drainage, and the more thoroughly the soil is prepared before planting, the longer they can remain undisturbed. Rhizomes of *C. caulescens* are more susceptible to rotting compared with the other species and require an especially well-drained site in order to flourish. For new clivia plantings, large quantities of well-decomposed compost to which bone meal has been added should be dug in to a depth of about 20 cm. A protected environment that is not subject to regular strong winds should be chosen for species with relatively soft-textured foliage, i.e. *C. caulescens, C. gardenii, C. miniata* and *C. robusta*, whereas coastal forms of *C. nobilis* with their somewhat harder, leathery leaves can withstand strong wind quite easily. Plant the vigorous, clump-forming types of *C. miniata* and the tall-growing *C. robusta* and *C. gardenii* about half a metre apart, while the slower-growing *C. caulescens* and smaller *C. nobilis* can be planted closer. Forms of *C. robusta* from swampy habitat can be planted in poorly drained areas of the garden like around shady ponds, while *C. miniata*, like *Agapanthus praecox,* is useful in stabilizing steep shaded banks.

Right: *Clivia caulescens* likes a protected environment, away from strong winds

Container subjects

All the *Clivia* species, particularly *Clivia miniata* make outstanding container subjects. They flower especially well when their roots are slightly restricted and, provided they are well fertilized, perform admirably in large terracotta and plastic pots, or large wine barrels. In addition, their attractive evergreen leaves and brightly coloured ripe berries provide interest throughout the year. *C. miniata* is particularly decorative when grown as a specimen plant in large terracotta pots flanking shady garden stairs or on either side of a shady front door. Clivias can also be grown most successfully as indoor plants in positions receiving good light but no direct sunlight. Once established, container-grown clivias can be left undisturbed for many years, until flowering performance diminishes. It is essential to provide excellent drainage for container-grown clivias as the roots will soon rot under soggy, poorly aerated conditions. There must be sufficient drainage holes at the bottom of the container, and a thick layer of broken crocks or stone chips should be placed over these.

Clivias can be grown in a number of different pot sizes, depending on the particular form. For larger forms of *C. caulescens, C. gardenii, C. miniata* and *C. robusta*, a 35 cm diam. plastic container is recommended, while for smaller forms of these species, as well as *C. mirabilis* and *C. nobilis*, a 30 cm diam. pot is suitable for the warm climates experienced in southern Africa, while in the cooler climate of the Northern Hemisphere, terracotta pots are preferable. Although *Clivia* roots generally spread out horizontally and not downwards, the longer the container, the faster the growing medium will drain. If the growing medium is sufficiently coarse and well drained, it is not necessary to specially place material like pine bark nuggets or broken crocks over the drainage holes at the bottom of the container. Re-pot in spring or early summer, as new growth begins, not in the heat of summer.

Growing medium

Clivias can be grown successfully in a wide variety of media, provided their three most important requirements are met: excellent drainage coupled with retention of sufficient moisture, excellent aeration and high humus content. *C. robusta* is exceptional in usually growing in seepage terrain in habitat, but does not require constantly wet conditions under cultivation. In the wild, the thick fleshy roots of clivias spread out horizontally in the decomposing leaf litter in which they grow and from which they derive their nutrients, or in the mossy layer covering boulders in the case of *C. caulescens*.

The water-retaining velamen outer layer covering the roots enables the plants to survive unfavourable conditions with ease. *Clivia* growers will discover their own preferred medium, based on local climatic conditions, and no rigid rules can be laid down in this regard. At Kirstenbosch the medium used for container plants consists of equal parts of well decomposed compost, sterilized, coarsest grade river sand and coarse, composted pine bark, while in the garden itself, large quantities of well decomposed compost are dug into fertile garden loam. Heavy clay soils should be avoided.

The 'river sand' available in many garden centres is often not really river sand in the strict sense, but sediment that is much too fine, and should be sieved in order to retain just the coarsest grains. Very good results can also be obtained using equal parts coarse, composted pine bark, leaf mould and sterilized coarse industrial (silica) sand. Another successful medium for use in containers is equal parts of acid peat and leaf mould. An orchid potting mix recommended for cymbidiums is also suitable for clivias, mixed equally with coarse river sand and acid compost. The ideal pH for 'green-leaf' clivias is slightly acid, from 5.5–6.5, which is roughly equivalent to that of leaf litter in the wild. For variegated clivias, a less acid or neutral pH ranging from 6.5–7 is preferable. In new gardens it is essential to remove as much builder's rubble as possible, as it may contain mortar, resulting in excess alkalinity of the soil.

The inclusion of material that allows for good soil aeration such as well decomposed compost, milled bark or coarse river sand is essential for container grown plants in order to prevent waterlogged soil and subsequent rotting. In coastal gardens, large quantities of decomposed compost or other organic matter should be added to very dry, sandy soils, and a layer of mulch can also be used to conserve moisture. (In the United Kingdom a suitable medium for container-grown clivias is equal parts of a loam-based potting compost like John Innes no. 2, with additional leaf mould and grit).

Best flowering results are obtained from container-grown clivias when their roots are somewhat restricted. Ideally, seedlings should be re-potted once

Below: An ideal growing medium for most container-grown clivias is equal parts of well decomposed compost, coarsest river sand and composted pine bark

every six months during their first and second years, then annually until they have reached flowering size, following which re-potting every second or third year is recommended. Failure to re-pot regularly results in the medium becoming compacted and consequently poorly aerated, which leads to root rot.

C. mirabilis and all variegated clivias have a greater susceptibility to fungal disease of the roots, and therefore a much coarser medium is recommended, consisting of eight parts coarse, washed river sand or grit, five parts well rotted composted pine bark, and five parts cymbidium orchid mix or charcoal.

Below: *Clivia miniata* 'Shima-fu'. Variegated clivias have a greater susceptibility to fungal disease of the roots and require a much coarser growing medium

Watering

Once established, clivias are remarkably drought resistant in all but the driest of conditions, provided they have sufficient shade and plenty of organic matter incorporated into the growing medium. Ideally, the five summer rainfall species (*C. caulescens, C. gardenii, C. miniata, C. nobilis* and *C. robusta*) should receive regular watering during their active growing period during the summer months, but less water during winter. *C. mirabilis*, the only winter rainfall clivia, is adapted to summer drought and thus requires most of its moisture from late autumn to early spring. Clivias easily survive long periods of drought even during their active growing period, but being such versatile plants they can, with the exception of *C. mirabilis*, easily withstand heavy winter rainfall, as is experienced in the southern suburbs of the Cape Peninsula, provided the growing medium is sufficiently well drained.

Watering frequency depends on a number of factors, including the age of the plants, prevailing weather conditions, and the type of container and growing medium being used. The older and more well established the plant, the less water it needs. In southern Africa, plants grown in terracotta pots dry out much more rapidly than those in plastic containers, and have to be watered more frequently. A suggested general watering programme for containerized, summer-rainfall clivias is a thorough drench once per week from early spring to late autumn for those in plastic containers, and twice to three times per week for those grown in terracotta pots. During winter, a drench once or twice per month is suggested. For *C. mirabilis*, whose roots are extremely susceptible to rotting, a drench once every two weeks is suggested during the winter growing period, reduced to just once per month in summer. Ideally, the growing medium for all clivias should reach a point of near-desiccation before the next drench is given, and it should never become waterlogged. In deciding whether to water or not, and depending on the depth of the container, the upper 5–10 cm of growing medium can be probed with the fingers; if it is bone dry to a depth of 5–10 cm, then the plant is ready to receive a drench, otherwise refrain from watering. Judging the level of moisture in a container (depending on the medium being used), can also be achieved simply by lifting the container; if it feels rather light, then watering is required, but if it feels heavy then it almost certainly is moist enough.

Watering should ideally be carried out in the early morning, allowing excess water enough time to dry by evening, thus reducing the incidence of fungal attack. As far as possible, refrain from watering directly into the crown of the plant (this is very important with mature plants of *C. mirabilis*) especially at the hottest time of year, but rather water in a circle some distance away from the crown, thus reducing the incidence of fungal and bacterial rotting.

Clivias don't require high humidity and normal watering is sufficient without wetting the leaves. A layer of dust often accumulates on the leaves of containerised clivias when grown on patios or under cover, and this can be sprayed off using a fine garden nozzle. Make sure containers are not placed in positions where they will overheat on very hot days. Water quality greatly affects the overall health of the *Clivia* plant, therefore use rainwater whenever possible.

For established plants in garden beds, a thorough drench once per week during the summer growing period should be sufficient, with only an occasional drench, approximately once per month during winter, in the absence of rain. Regular fertilization of container-grown clivias often leads to a build-up of salts, and it is important to leach the soil about once per month by drenching with pure water in order to leach excess salts out of the growing medium.

In conclusion, it is preferable to err on the side of dryness than to be too wet. The roots of variegated clivias are especially susceptible to rotting if overwatered, and extra care should be taken with these special plants to ensure their growing medium drains sufficiently, and that it reaches a point of near desiccation before the next drenching is given.

The roots of *Clivia mirabilis* require infrequent watering as they are extremely sensitive to fungal rotting

Feeding

Clivias are gross feeders, and the quantity and quality of flowers, leaves and fruits can be greatly enhanced through regular feeding via both the foliage and roots. Clivias grown in media containing sufficient compost/humus will grow very well, but correct feeding practises will enhance their performance. Feeding is especially important when plants are grown in inert media such as milled bark, in which nutrient availability is low. Bear in mind, however, that fertilizers high in nitrogen tend to cause excessive leaf growth at the expence of flowers. Young seedlings respond very well to liquid feeds of seaweed extract like Seagro and Kelpak, as well as to Supranure, which contains a growth stimulant. Apply every three or four weeks, either as a foliar spray or as a soil drench. For mature plants of the five summer-growing species (*C. caulescens, C. gardenii, C. miniata, C. nobilis* and *C. robusta*) grown in the garden or containers, applications of a granular, complete fertilizer, relatively low in nitrogen, low in phosphorus but high in potassium, such as 3:1:5 (expressed in order of nitrogen, phosphorus and potassium) is recommended. Apply three times during the summer growing season, in early spring, early summer and late summer. Nitrogen stimulates leaf growth, phosphorus develops a good root system and potassium regulates the growth of stems. For the winter-growing *C. mirabilis*, a single application in late autumn is sufficient. An additional trace nutrient element fertilizer like Trelmix is recommended for all clivias if plants show deficiency. Nitrogen (N) deficiency is seen as general yellowing of the leaves, first observed in the oldest leaves, and stunting of new leaves; phosphorus (P) deficiency is recognized as yellowing or blue-green or purple colours in leaves, first seen in older leaves, and in potassium (K, potash) deficiency, leaves develop scorched margins, also first seen in older leaves. Osmocote High K is a recommended slow-release fertilizer which fertilizes the soil over a long period, stimulating good scape length and inflorescence size. It is best applied in early summer and early autumn. An application of a seaweed extract fertilizer when the buds form has been found to ensure longer flower stems. In addition to applications of granular fertilizer at certain times of the year, all container-grown clivias benefit from monthly applications of liquid feed in the form of a foliar spray such as Nitrosol or Kelpak. The application of the non-toxic, non-burning, organic fertilizer 'Neutrog Bounce Back' (derived from chicken litter) gives excellent results when mixed into the growing medium or sprinkled liberally on the surface. It provides water insoluble nitrogen that is released over a long period, and supplies a full balance of nutrients that do not leach rapidly.

It is important to note that a build-up of salts, especially in containers, may occur as a result of regular feeding of clivias, and can damage the roots. It is best to flush out these excess salts once per month by watering only with pure water, preferably rainwater, as water quality varies from time to time and may result in additional salt accumulation. Salt accumulation is seen as a white or brownish-yellow deposit on the soil, or as a crust against the inside rim of the container.

Deficiency in potash is thought to be a possible cause of failure of the scape to elongate. Feeding the plants with

3:1:5 after flowering and during seed formation will result in larger, healthier seeds that mature faster. For established *Clivia* plantings in the garden, a generous mulch of well decomposed top-dressing compost applied in autumn and again in spring, together with an organic fertilizer like 'Neutrog Bounce Back' or bone meal delivers excellent results, and being shallow-rooted plants that resent disturbance, there is no need to dig-in compost.

It is essential not to over-feed variegated plants, as this stimulates the chloroplasts of yellow areas of the leaf to turn green, minimizing the variegated effect.

Below: *Clivia miniata* hybrid raised at Wisley Gardens, England. All summer-growing clivias are gross feeders and benefit greatly from regular applications of fertilizer

Below: Ripe *Clivia miniata* berry containing mature seeds

Opposite: Young *Clivia miniata* seedlings

PROPAGATION

Propagation by seed, separation of offsets and stolons are currently the most widely used methods of increasing stocks of *Clivia* plants. Tissue culture is a recent new method that has been successfully employed, but much research has still to be done in order to make this method commercially viable.

Seed

Propagation of clivias by means of seed is an easy, inexpensive way of increasing stocks. The seeds can be harvested at any time once the berries have started to change colour to deep red, bright red-orange or pale to dark yellow, and they are best sown as soon as possible after harvesting.

Harvesting seeds from berries that have been attached to the pedicels for extended periods must be undertaken with great care as the seeds often begin to germinate while still inside the berry, and the developing radicle can easily be damaged. If not harvesting seeds, the berries can simply be left on the plant to fall off naturally, and the seedlings will establish themselves next to the mother plant. The pericarp of berries that have been harvested but not had their seeds removed, eventually shrinks, loses colour, and becomes very hard, yet the seeds within can remain viable for many months under these conditions.

The berries of most species and hybrids ripen from mid to late winter, whereas those of *C. mirabilis* ripen from mid to late autumn. To harvest seeds from newly ripe berries, break open the outer fleshy layer of each berry, remove the seeds and clean them by rubbing off the sticky outer membrane that envelops each seed in a bowl of lukewarm water to which a squirt of dishwashing liquid has been added. Allow the seeds to dry for about 30 minutes, then let them stand in a mild fungicide solution like captab (e.g. Kaptan) or mancozeb (e.g. Dithane) for 20 minutes and allow to dry. Alternatively, they can be placed in a plastic bag and shaken-up with captab or mancozeb powder prior to sowing. The seeds should be sown as soon as possible thereafter as germination will be delayed if they dry out too much.

Seeds are best sown in deep seed trays in a well drained, well aerated medium such as one part coarse river sand to three parts finely milled, composted pine bark or decomposed pine needles. Press each seed into the medium so that it rests just below the surface, allowing about 20 mm between each seed to provide adequate room for the developing seedling.

Placing the seeds on top of the surface is not recommended as they frequently become dislodged when watered, and the developing radicle has difficulty in penetrating the medium (Duncan 2004b). Place the seed trays in a warm, shaded, well aerated position under cover, and keep moist by watering thoroughly once or twice per week with a fine rose, preferably with water that has been stored at room temperature.

Ideal temperatures under which to sow seeds is 15–21 °C, but seeds sown under cold conditions will simply remain dormant until conditions warm up, before germinating. Alternatively, seeds can be stored until temperatures rise in early spring. Germination takes place within 4–8 weeks, with the appearance of a radicle, followed by the first cotyledon (seedling leaf). The radicle becomes the primary root of the plant and develops a circle of root hairs just below its tip. Eventually the primary root withers and is replaced by adventitious roots that become the adult roots of the plant. Once germinated, seedlings of *C. caulescens, C. gardenii, C. miniata, C. mirabilis* and *C. robusta* develop rapidly under ideal conditions, while those of *C. nobilis* are generally much slower.

A series of experiments carried out in 1999 at the Seed Laboratory at the Kirstenbosch Research Centre found that pre-cleaned seeds stored at 5°C gave a high percentage germination even after one year, whereas seeds stored at the same temperature in their fruits resulted in a high germination rate for up to six months; thereafter a much lower rate of germination was obtained because they became heavily infected with fungi and tended to rot in storage (Brown & Prosch 2000).

Seedlings benefit from a high nitrogen, weekly application of Kelpak or Supranure that stimulate rapid vegetative growth. (These products are too nitrogen-rich for mature plants and discourage flowering). Once seedlings have developed two leaves, a light application of Neutrog Bounce Back can be made to the soil surface.

Opposite: The berries of most *Clivia* species and hybrids such as *C. miniata* Pat's Gold strain ripen from mid to late winter

Left: A tray of two-year old *Clivia miniata* seedlings ready for planting out

After seedlings have reached 12 months, they can be potted-up singly into 15 cm diam. pots or 2-pint plastic bags. In the spring of their second year, plant them into 20 cm diam. pots or 1-pound plastic bags, and allow to grow-on for a further year. Depending on their progress, at the beginning of their third or fourth season they can be planted out into the garden or into permanent containers. The roots should be slightly constricted before young plants are potted-on.

Although certain dwarf *C. miniata* hybrids currently being produced commercially in Belgium can be induced to flower in as little as two years from seed, *C. miniata* seedlings generally take at least three or four years to flower for the first time under ideal conditions, as do those of *C. caulescens*, *C. gardenii* and *C. robusta*.

It is estimated that *C. mirabilis* may flower in less than five years, but seedlings of *C. nobilis* are generally rather slow-growing, and take at least five years or more to flower. Considerable variation may occur in the time taken for seedlings of *C. miniata* to flower from seed. Under ideal conditions some may flower in as little as 24 months from sowing for certain dwarf Belgian strains, while others may take up to five years or more. In experiments carried out in Belgium, where seed-grown plants are raised in frost-free greenhouses, it was found that flower bud initiation took place for the first time after 12–13 leaves had formed. For scape elongation, a cold treatment of several weeks is required. The successful commercial production of clivias as flowering pot plants hinges on the minimum time needed to grow the plants to flowering size, and

Shigetaka Sasaki

homogenous flowering. In increasing the growing temperature, vegetative growth is accelerated and plants can be brought into flower in two years, but a cold period of several weeks is needed to get the scape to elongate sufficiently above the leaves. Once the desired stem length has been achieved, increasing the temperature and providing extra light hastens flower development in order to produce the product before Christmas (Van Huylenbroeck 1998).

When propagating *C. miniata* var. *citrina* and pastel forms of *C. miniata* var. *miniata* from seed, it is important to note that seeds harvested from yellow- or pastel-flowered parents will not necessarily all yield yellow-flowered plants unless the parent plants are a true-breeding strain.

It is possible to determine at an early stage whether seedlings of *C. miniata* var. *citrina* will produce yellow or pastel flowers by examining the base of the first leaves produced by the seedling at any stage during the first or second year of growth: if the colour of the base of the leaves is plain green (unpigmented), the plant will usually produce yellow or pastel flowers, but if the base of the leaves is dark reddish-maroon, the flowers will be in shades of orange or red.

Growing variegated clivias from seed

Occasionally, a number of seedlings with limited amounts of chlorophyll, as well as some with no chlorophyll in their leaves may appear out of a batch of otherwise healthy seedlings. Those with limited chlorophyll are referred to as being variegated, while those devoid of

Opposite: *Clivia miniata* Daruma strains can be raised from seed to flowering within three years

Left: Basal leaf pigmentation in *Clivia miniata;* at left, an unpigmented seedling indicating a plant that will produce yellow flowers; at right, purplish-maroon pigmentation indicating a plant that will produce orange or red flowers

Below: *Clivia miniata* 'Akebono-fu' seedlings

chlorophyll are referred to as albinistic, an inherited, lethal condition that always leads to the death of the seedlings. With special care, variegated seedlings can usually be persuaded to continue growing, but inevitably albinistic seedlings die once the food reserves in the seed endosperm have been used up. This genetic condition is usually brought about by inbreeding and is especially common in seedlings of yellow-flowered and variegated plants of *C. miniata*, and to a lesser extent in orange-flowered plants, but rarely in other *Clivia* species. Being much weaker than uniformly green-leafed clivias, variegated plants are much more difficult to raise from seed, and their germination rate is usually rather low (30% or less). Albinistic seedlings are doomed and should be discarded as soon as they are noticed.

The sowing medium for variegated clivias should be particularly well drained and well aerated, with a greater percentage of coarse sand incorporated into the medium in order to reduce the incidence of fungal disease of the roots, to which they are highly susceptible. A recommended sowing medium is three parts coarse river sand and one part finely milled, composted pine bark or well decomposed pine needles. If possible, the sand component should be sterilized, which is easily achieved by treating it with boiling water, then allowing it to dry for a few days before making-up the sowing medium.

The seeds are sown in the normal manner, placed in a shaded, well ventilated but protected position, and irrigated with water stored at room temperature. Like those of ordinary clivias, fresh seeds can take up to two months to germinate. Being rather weak, seedlings of variegated clivias are not able to tolerate strong light. When exposed to direct sun, the yellow and white areas of the leaves become severely scorched, and must be kept well shaded at all times. They also cannot tolerate hot or humid conditions and must have excellent ventilation at all times.

Once they have developed a strong leaf within three months, they can be potted-up individually into 10 cm diam. pots and after nine months, into 15 cm diam. pots. In the spring of their second year, they can be potted-up into 20 cm pots, and at the beginning of their third season they can be planted into permanent containers. They generally flower in their third or fourth season, once they have about 14 leaves.

Variegated clivias are only suited to container cultivation as their roots are too susceptible to fungal disease to stand up to the rigours of garden cultivation. When applying fertilizers to variegated plants it is best not to apply it directly onto the leaves as they may burn. It is safer not to water plants directly onto the crown but some distance away. The better the water quality, the healthier the plants will be. The most frequent disease of variegated clivias is root-rot due to insufficiently well-aerated growing media. A close watch must be kept on the condition of variegated plants in order to react to their needs timeously before major damage is done.

Pollinating clivias

There are several methods by which *Clivia* flowers can be artificially pollinated, and each grower will discover his or her most convenient one. Before commencing with pollination, the anthers must be ripe: they must have split open and the pollen should be loose and easily removed. Similarly, the stigmas of the flowers one intends pollinating must be receptive: the stigmatic branches must have separated into a star-like position, and should be slightly moist and sticky. Pollen can be collected by dabbing the base of a fine water paint brush over the ripe anthers, then dabbing the brush over a receptive stigma.

Alternatively, one can slightly dampen the tip of an ear bud and transfer pollen in the same manner, or cut off and hold a stamen with a pair of tweezers and brush the anther over a stigma. Yet another method is to use one's fingers by lightly pinching an anther between the thumb and index finger until the pollen adheres, then pinching the stigma very lightly until the pollen adheres. Be sure to use different brushes and ear buds when pollinating different plants, and remove pollen from fingers and hands by thorough washing before proceeding with pollination of the next plant.

The best time to pollinate is in the early morning, while the fluid on the stigmatic branches is still sticky and moist. As the flowers open on different days, one should make sure that each flower is ready for pollination before it is carried out, and each flower should be pollinated several times over a period of days to ensure that pollination does take place.

Opposite: Exposure to direct sun may result in leaves of variegated clivias being scorched

To self-pollinate the flowers of a particular plant, isolate it effectively before the stigma ripens so that it cannot be pollinated by wind, insects or sunbirds. When cross-pollinating, to ensure that pollen from one flower does not land on its own stigma, emasculate the flower by cutting off the anthers well before they ripen. In order to reduce the occurrence of albinism, that inevitably causes the seedlings to die, when pollinating variegated clivias, it is important to only pollinate flowers carried on non-variegated pedicels, or those with very light variegation carried on the green or less variegated side of the scape. Pollen can be stored for up to two months by placing anthers in small glass vials or plastic capsules into a larger airtight container, together with a sachet of drying agent such as silica gel, into the vegetable compartment of a fridge for use during the current flowering season. It can also be stored in the freezer compartment for use up to three years later (Lötter 1998).

It is of the utmost importance to keep detailed records of the crosses one has made, and each pollinated flower should be tagged around its pedicel with details of the cross, on a pencil-written label.

Division

Division of clivias by separation of offset material is the most reliable method of obtaining plants that are exactly true to type. They can be divided at any time of year in the warmer parts of South Africa like the coastal areas of KwaZulu-Natal and the Lowveld, but in other areas like the Western Cape the best time to divide the summer-growing species is in spring and early summer, as new vegetative growth begins immediately after the flowering period. The autumn- and winter-flowering *C. gardenii* and *C. robusta* are summer growing and can also be divided at this time. Plants of the winter-growing *C. mirabilis* occasionally produce offsets and these should be separated in autumn, as active growth begins. All faded flower heads should be cut off to prevent the formation of seeds while the newly divided plants establish themselves.

Dig up large clumps in garden beds, then shake off as much soil as possible and gradually prize the plants apart by placing two forks back to back in the centre. A sharp spade can be used to cut off individual plants. Offsets should not be forcibly broken off the mother plant but removed by persistent tugging; if they do not give way, they are not ready to be separated. Alternatively, they can be cut off using a clean, sharp knife, ensuring that as much root material as possible is retained. Some of the outer leaves can be removed to reduce water loss, but the leaves should not be cut down as is often the practice when dividing evergreen agapanthuses. Cut or damaged rhizome surfaces should be cleaned, then dusted with captab (e.g. Kaptan) or 'flowers of sulphur'. Separated offsets should ideally be re-planted as soon as possible in damp (not wet), well composted soil and kept well shaded. Newly divided clivias whose roots have been substantially reduced will usually not flower for a year or two, until they have become well established. While most forms of *C. miniata* and certain forms of *C. nobilis* reproduce rapidly by offsets, *C. caulescens, C. gardenii* and *C. robusta* are slower to reproduce by this method. Certain clones of *C. gardenii* as well as *C. miniata* var. *citrina*, especially 'Natal Yellow' reproduce vigorously by the formation of subterranean stolons produced from the top of the rhizome, each of which develops into a new plant some distance away from the mother plant, and these are very easily separated from the mother plant when large enough.

While vegetative propagation from uniformly green-leafed mother plants yields identical progeny, this is not the case with variegated plants. In the latter, an offset can look very different to the mother plant, depending on which side of the plant the division develops. If it develops on the side with broad bands of yellow or white it will look similar, but if from the side without high incidence of variegation, the progeny may be uniformly green, or have very few stripes.

Below: A thick clump of *Clivia miniata* var. *citrina* ready for division

Tissue culture

Clivia miniata has been successfully cultured using fruit and floral explants, and the fruit wall (pericarp) is the most successful explant for *in vitro* multiplication of *Clivia* (Finnie 1998). However, plantlet regeneration is very slow, and is currently not an economically viable method of propagation. Two drawbacks to this method are that the fruit wall is only available seasonally, and the ideal time for taking explants is very specific, with no easy way to determine the optimum time. Miyoshi & Company became the first commercial concern to mass produce clivias by tissue culture, which they did with the hybrid *C. miniata* 'Vico Yellow' (see pages 118, 121).

Below, left: Snails can do great damage to *Clivia* flowers, fruits and scapes.

Below, right: Sun scorch on a leaf of *Clivia miniata*

Opposite: Lily borer (amaryllis caterpillar) devastates *Clivia* plants

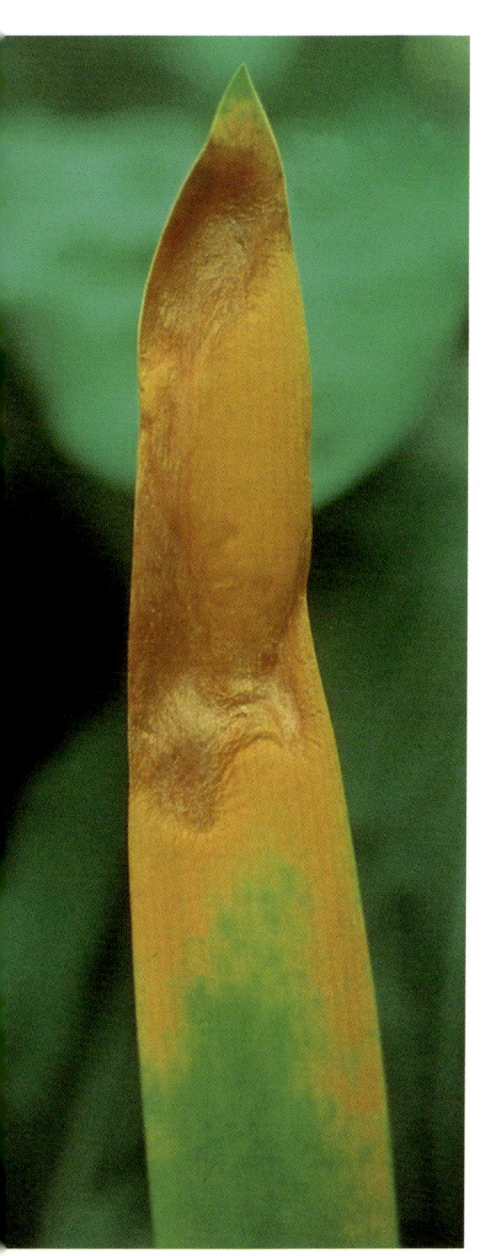

PESTS AND DISEASES

Despite the general ease of cultivation of most clivias (with the exception of *C. mirabilis*), they are subject to a fairly wide range of pests and diseases, and the following measures are suggested for their control. Unfortunately, the identification and treatment of a number of fungal diseases of clivias is still in its infancy, but where the latter is suspected, an immediate course of action is simply to reduce moisture and, if possible, improve air circulation, and in more instances than not, the problem will resolve. Once a pest or disease has been identified, it is best not to delay in taking remedial action, as the longer the problem persists, the more difficult it becomes to treat. For example, treat mealy bug and aphid infestation immediately it is noticed, thus reducing the risk of viral infection. Wherever possible, environmentally friendly methods are recommended, but in many instances the use of chemicals is necessary. It should be borne in mind that certain insect pests such as mealy bug and red spider sometimes become resistant to continued use of the same product, therefore different products should be alternated in order to prevent the incidence of resistance. Ensure that the undersides of the leaves are well covered with the substance being applied, as this is where some pests hide and multiply. Read the instructions very carefully before commencing with application of pesticides and fungicides, and do not exceed the recommended dosage as this may result in burning of the leaves. In general, pesticide and fungicide 'cocktails' should not be used as different chemical substances may not be compatible and may cause unnecessary damage to the plants. The instructions provided with some pesticides and fungicides indicate which other substances they are compatible with. A spreader/sticker can be added to certain pesticides and fungicides in order to prevent the spray being washed off the leaves, but in some instances this is contra-indicated, such as with the systemic fungicide Funginex. Alternatively, 5 ml of dishwashing liquid can be added to 5 ml of any pesticide or fungicide spray.

The more stressed a plant is, the more susceptible it is to pest infestation and disease. Knowledge of appropriate cultivation practises can minimize the occurrence of pest and disease attack in container-grown clivias, including growing plants in appropriately sized containers, promptly removing dead or dying outer leaves and disposing of them by burning or burying, avoiding fertilizers high in nitrogen, ensuring adequate ventilation at all times, spacing plants adequately and discarding, or at least isolating, suspected virus-infected material immediately.

Right: Leaf miner damage on a *Clivia* leaf
Opposite: Lily borer caterpillar on *Clivia nobilis* (left) and leaf damage (right)

Pests

Aphids These small soft-bodied, sap-sucking insects (also known as greenfly or blackfly) with relatively long legs and antennae mainly attack developing flower buds in *Clivia*, and to a lesser extent, developing leaves, causing deformities. They are one of the principal vectors of insect-borne viral disease. Like mealy bug, they excrete sticky honeydew that coats the foliage, which becomes infected with fungal sooty moulds, resulting in leaves eventually turning yellow and dropping off. For large numbers of plants, spray with mineral oil (e.g. Oleum), or drench the soil with imidacloprid (e.g. Kohinor 350 SC). Individual plants are effectively treated using an aerosol spray containing tetramethrin (e.g. Wonder Garden Gun).

Gall midge fly The small yellow larvae of this fly occasionally burrow into young *Clivia* flower buds, causing malformation, and are also responsible for secondary bacterial and fungal infection.

Preventative spraying is necessary in susceptible areas as soon as buds begin to appear. Spray with partially environmentally compatible fenthion (e.g. Lebaycid) as a full cover spray.

Leaf miner Like the lily borer moth, the leaf miner moth lays its eggs on the undersides of *Clivia* leaves (except those of *C. mirabilis* and *C. nobilis* which are too leathery) and the tiny larvae live inside the leaf, usually in the upper parts, creating intricate pale brown patterns as they feed.

The larvae do not bore into the rootstock like the lily borer, but cause the leaves to turn brown and die back from the tips. They are present mainly in summer, and in areas prone to severe infestation are best combatted by spraying preventatively with a carbaryl-based insecticide such as Carbaryl or Karbaspray, or dusting with Karbadust, which are partially environmentally compatible.

Lily borer (also known as amaryllis caterpillar) This is without doubt the most important pest known to *Clivia* growers in southern Africa. Known scientifically as *Brithys pancratii,* it is widely distributed in Africa and southern Europe (Annecke & Moran 1982). The night-flying moth usually lays 50–100 eggs on the undersides of the *Clivia* leaf, from spring to early winter. The tough leathery leaves of *C. mirabilis* and *C. nobilis* are seldom attacked but the four remaining species which have relatively soft, easily penetrable leaves are highly susceptible, especially those of *C. caulescens* and *C. miniata*. After hatching, the young caterpillars rapidly bore into the leaf tissue which they proceed to consume in vast quantities, causing it to turn black, while moving towards the base of the leaf. Typically, by the time this pest is noticed, a great deal of damage has already occurred. If left unchecked, the caterpillars proceed towards the leaf bases, eventually destroying the growing shoot and even causing the death of the

159

plant. Large caterpillars can be picked off by hand, or, affected leaves can be cut off, stamped on and placed on the compost heap. This method of control has limited success as inevitably some caterpillars escape one's attention. A much more effective method of control is to spray or dust affected plants immediately with a carbaryl-based insecticide such as Carbaryl, Karbaspray or Karbadust, which is partially environmentally compatible. Indicator plants for the presence of lily borer in the garden include the amaryllids *Ammocharis coranica*, *Cyrtanthus elatus* (*Vallota speciosa*), *Crinum moorei* and *Nerine bowdenii*, whose soft fleshy leaves are the first to be targeted. Lily borers are at their most active during the hot summer and early autumn months, and in heavily susceptible areas, plants should be sprayed preventatively.

Below, left: A cluster of lily borer eggs on the underside of a *Clivia* leaf

Below, centre: Mealy bug infestation on young *Clivia* leaves

Locusts These are dimorphic grasshoppers which, in their adolescent phase behave like other grasshoppers, but in the gregarious phase they form swarms. They are only a problem in some summer rainfall parts of South Africa, such as in Mpumalanga, and can do great damage to *Clivia* foliage. The adolescent phase is the most dangerous one when most damage is done, and can be controlled either by catching them by hand or spraying or dusting plants preventatively with a carbaryl-based insecticide (e.g. Carbaryl, Karbaspray or Karbadust).

Maggots Watery blemishes develop under the pericarp (outer skin) of the ripe berries, turning it transluscent brown as the maggots feed on the inner pulpy layer. Although resulting in unsightly berries, the maggots do not damage the seeds.

Below, right: Blemished *Clivia* berries (top) caused by maggot infestation (below)

Mealy bug These small white, waxy sucking insects attack the leaf bases, upper and lower surfaces of all *Clivia* species, especially *C. miniata*, eventually causing it to turn yellow, and can also transmit viral disease. Their presence should immediately be suspected when oily white deposits are noted on the leaf bases, especially on the undersides, and they are prevalent on clivias grown in containers under enclosed, warm and dry conditions. They are spread from one plant to another mainly by Argentine ants that feed on the honeydew secreted by the mealy bugs. In return, the ants protect the mealy bugs from their beetle predators. A black mould may sometimes develop on the honeydew excretion.

Ants can be controlled in an environmentally friendly way by pouring neat Jeyes Fluid down ant holes, then washing it down with water. The process should be repeated frequently. For infected clivias, drench the soil with imidacloprid (e.g. the contact systemic insecticide Kohinor 350 SC) for container-grown plants, which provides protection for up to one year. It is said to have minimal or no impact on beneficial insects and earthworms, but should however be handled with great care, as with all insecticides. The recommended dosage should not be exceeded as this rapidly causes burning of the leaf tips. Mineral oil (e.g. Oleum) and the contact insecticide chlorpyrifos (e.g. Chlorpirifos) can also be used to treat infected plants.

Mealy bugs eventually become resistant to continued applications of chlorpyrifos, and Kohinor 350 SC is by far the better product to use, although it is not yet readily available from garden centres in South Africa.

Mole rats In some parts of South Africa such as in the Eastern Cape and KwaZulu-Natal, mole rats can be troublesome in that they consume the fleshy roots and rhizome of the *Clivia* plant. A temporary measure of control can be achieved by planting clivias into strong, sunken wire baskets, and placing rocks around each plant. Alternatively, sow seeds of the annual, *Euphorbia lathyris*, commonly known as the caper spurge or mole plant. When the caper spurge seedlings have reached 15 cm high, they can be transplanted 4 m apart in areas where moles are active. They grow to 750 mm high, and once the roots are established the moles leave the area. This erect, unbranched biennial with leathery grey-green leaves, produces terminal umbels of yellow flowers, enclosed by bright green bracts, followed by caper-like fruits. They die off at the end of their second year, and should be sown again if they have not reseeded of their own accord. Ensuring that there are always plants in affected areas will keep the moles away.

Red spider mite These tiny red, sap-sucking insects sometimes occur on the undersides of *Clivia* leaves, and are prevalent in warm, enclosed conditions. Their presence is recognized by a fine greyish web, and can be controlled by spraying with mineral oil (e.g. Oleum) or drenching the growing medium with imidacloprid (e.g. Kohinor 350 SC), a contact systemic insecticide.

Rodents Rats, mice and tree squirrels are of considerable nuisance value in that they feed on the pericarp (outer skin) and inner pulp of ripe *Clivia* berries in late winter in various parts of southern Africa, but leave the seeds undamaged. For control of rats and mice, use environmentally friendly

Alpha cellulose (e.g. Eco Rat pellets), an organic formulation. Squirrels are sensitive to uric acid and may be kept at bay by thinly scattering guano near mature plants – if you can stand the smell.

Scale This pest is recognized as soft or hard, pale brown or reddish small raised shells on the scapes and upper and lower leaf surfaces. Like mealy bugs, it eventually causes the leaves to turn yellow prematurely, desiccate and fall off. It is prevalent in warm, poorly ventilated conditions and is best treated by drenching the soil with imidacloprid (e.g. Kohinor 350 SC), a contact systemic insecticide that provides protection for up to one year.

Slugs and snails These can do great damage to *Clivia* flowers, developing flower buds, the scape and fruits, especially the common brown garden snail, also known as the European brown snail, *Helix aspersa*. They feed on leaves to a much lesser extent but may at times even feed on velamen, the water- and nutrient-retaining corky outer layer of exposed *Clivia* roots. They are also transmitters of viral disease, to which all clivias are susceptible. Telltale evidence of their nocturnal forays is often seen in their sticky, slimy paths left behind. Pick off the culprits by hand, or, in severe infestations of *Clivia* plantings in the garden, keep ducks to do the job for you (Muscovies or Dutch quackers are ideal). Alternatively, place broken eggshells around the base of the plants, or make use of a recently introduced organic product in the form of iron phosphate pellets (e.g. Biogrow Ferramol) which can be scattered in the evening around the outside of pots and under raised benches. Multiple applications should be made in susceptible areas, especially following heavy rainfall. Ingestion of even small quantities of the bait causes slugs and snails to cease feeding, become less mobile and die within a few days. Pellets that are not consumed degrade and become part of the soil. Favourite daytime hiding places for inactive snails is under the rims of plastic containers, on the undersides of leaves of evergreen agapanthus plants, inside large dry, curled-up fallen leaves and frequently inside the open flowers of *C. miniata*.

Snout beetle These very destructive, small, grey or pale brown beetles are nocturnal and feed on *Clivia* flowers as well as their leaf blades and margins, leaving behind characteristic round bite marks. They are especially prevalent in the winter rainfall region of the Western Cape. During the day the beetles hide among dry leaves near the base of the plant, at the bases of the leaves or deep inside the flowers.

Below: Snail damage to *Clivia miniata* flowers

They can be picked off by hand at night with a bright torch, but the slightest movement causes the beetles to drop to the ground where they are completely camouflaged; it is best to place a cupped hand or container underneath the plant and shake the culprits off, then crush their hard exteriors. In severe infestation, spray with a cypermethrin-based insecticide (e.g. Garden Ripcord), which is partially environmentally compatible, as a full cover spray.

Thrips These minute, elongated, black or brown insects feed on the undersides of *Clivia* leaves, leaving behind characteristic white streaks, and also attack developing flower buds of late-flowering clivias, resulting in deformed flowers. They are prevalent during the hot summer months in the Southern Hemisphere and are transmitters of viral disease. Drenching the soil with the contact systemic insecticide imidacloprid (e.g. Kohinor 350 SC) will keep them away, or alternatively for preventative treatment of small numbers of plants, a light spray with the contact aerosol Wonder Garden Gun is highly effective.

Whitefly These small white flies are a problem in that they suck the sap of soft-leafed clivias, especially *C. miniata*, resulting in the leaves turning yellow and eventually falling off. They are active mainly in summer, and are found on the undersides of the leaves, where they multiply rapidly, flying up in large 'clouds' of individuals when disturbed. They are prevalent in warm, protected conditions, especially shadehouses and hothouses. For large numbers of affected plants, spray with cypermethrin (e.g. Garden Ripcord), or for individual plants, use tetramethrin in the form of an aerosol spray (e.g. Wonder Garden Gun). This pest should be sprayed every 3-4 days until its extremely rapid life cycle has been completely disrupted. Whitefly are partial to the leaves of numerous weed species, especially the undersides of thistle leaves; keeping weeds well under control will lessen their occurrence.

Below: Snout beetle damage to leaves and flowers of *Clivia miniata*

Diseases and nutrient deficiencies

Agapanthus fungus (leaf die-back)

An unidentified fungus (possibly *Sphaeropsis agapanthi*, previously incorrectly known as *Macrophoma agapanthi*) also attacks the leaves of clivias, mainly *C. miniata*, causing them to turn brown, dry out and die back from the tips, sometimes resulting in severe disfiguration of the plants. The affected parts become very dry and brittle to the touch. The disease is especially prevalent during the hot summer months, and one of the causes may be the result of an old growing medium that has become excessively compacted over many years as a result of decomposition, or having remained excessively wet for long periods of time, resulting in insufficient oxygen uptake by the roots. Certain clones such as *C. miniata* var. *citrina* 'Kirstenbosch Yellow' are especially susceptible to it. Remove badly affected leaves by tugging or breaking them off at the base of the plant, then re-pot the plants into a new growing medium and spray the whole plant preventatively with chlorothalonil (e.g. Bravo), the environmentally compatible, broad spectrum mancozeb (e.g. Dithane M45), or copper oxychloride (e.g. Virikop) as a full cover spray, ensuring that both surfaces of the leaves are well covered. Bravo is especially effective as it is not easily washed off leaf surfaces once spray residue has dried on the leaf.

Chlorosis This is the condition in which *Clivia* leaves prematurely turn yellow, white or pale green, due to a complete or partial loss of chlorophyll. It can be caused by inadequate or excessive light, or mineral deficiency. Mineral deficiency in alkaline soils is usually due to insufficient manganese or iron, which can be corrected by applying trace (micro) elements (e.g. Trelmix Trace Element Solution) and iron chelate, and mulching with acid compost. Builder's rubble is a source of excessive alkalinity in the soil and as much of it as possible should be removed. In acid soils, chlorosis is usually due to magnesium deficiency, which can be corrected by the application of magnesium sulphate (Epsom salts) as a soil drench, or trace elements (e.g. Trelmix Trace Element Solution).

Damping-off fungi These highly destructive soil-borne fungi (especially *Fusarium* and *Pythium*) are prevalent in seedlings of all *Clivia* species, although they can also attack adult plants, causing the leaves to die back rapidly from the tips or to rot off just below soil level. They are prevalent in humid, poorly aerated conditions and are best controlled by drenching the soil preventatively with mancozeb (e.g. Dithane M45), or by dusting the seeds with copper oxychloride (e.g. Virikop).

Opposite: Leaf die-back (left) and fungal leaf spots in *Clivia miniata* (right)

Below: Chlorosis (left) and flower stalk malformation (right) in *Clivia miniata* var. *citrina*

Flower stalk malformation This condition affects the scape (flower stem) of all colour forms of *C. miniata,* causing it to fail to elongate properly. It is a somewhat erratic occurrence and is thought to be due to potash (K) deficiency and/or failure to provide plants with a dry winter rest period. Treat affected plants by refraining from feeding or over-watering during the winter months, and by applying a basic granular fertilizer like 3:1:5 in early spring, early summer and late summer, as well as trace (micro) elements like Trelmix Trace Element Solution, at any time of year.

Leaf spots The foliage of *C. miniata* is subject to several fungal diseases which result in the formation of unsightly spots or blotches. Unfortunately very little is known about these diseases at present. The best course of action is to cut off and destroy affected portions of leaves as soon as large spots or blotches are noticed, improve ventilation, reduce humidity, ensure excellent drainage of the growing medium and treat affected plants with a systemic fungicide containing benomyl (e.g. Benlate), as a soil drench. Highly toxic fungicides like benomyl should only be applied in severe infestations, and under carefully controlled conditions.

Parasitic nematodes There are more than 2000 species of these microscopic, worm-like organisms, most of which live in soil and are invisible to the human eye. Certain nematode species feed on the sap of *Clivia* roots, resulting in characteristic swollen lumps or knots on the roots, and die-back of the leaf tips. Most chemical control products for nematodes are highly toxic but nematodes affecting clivias can fortunately be controlled biologically by fungi and other nematode species found in organic matter, that are their natural enemies. To this end, an effective biological control product called 'Plplus' containing the active ingredient fungus *Paecilomyces lilacinus* Strain 251, has been developed (Chubb 2002). It is added to the growing medium in container-grown plants as well as to soil in garden beds. Growing media infected by nematodes should never be re-used, but disposed of.

***Rhizoctonia* fungus** The symptoms of this fungus are similar to those of soft crown rot (except that it is slower to develop and is not accompanied by the strong rotting stench), causing discolouration and wilting of the foliage, resulting in the plant rotting-off at ground level and falling over. It is prevalent in *Clivia* seedlings as well as mature plants, especially under unhygienic conditions such as waterlogged growing media and poorly ventilated growing conditions. Lift affected plants immediately, remove and destroy the infected parts, then re-plant in a new medium and drench the roots with benomyl (e.g. Benlate). Alternatively, take preventative action by spraying with mancozeb (e.g. Dithane M45).

Rust fungi These regularly attack the foliage of *C. miniata*, especially variegated forms of this species, and to a lesser extent, *C. gardenii* and *C. robusta*. They are visible as sporadic hard, large green or red pustules, or dense concentrations of small reddish-brown pustules on the upper and lower surfaces of leaves which break open and shed spores, and are prevalent in poorly aerated, enclosed conditions. Improved ventilation helps alleviate the incidence of these fungi and in susceptible areas spray preventatively with environmentally compatible mancozeb (e.g. Dithane M45) as a full cover spray at weekly intervals until the disease is controlled, or with a systemic fungicide such as triforine (e.g. Funginex), which has a low toxicity to wildlife.

Below: Death of a *Clivia* plant due to *Rhizoctonia* fungus

Most systemic fungicides move upwards in plant tissue to areas of new growth, or from older to new leaves. The efficacy of some fungal sprays is greatly increased by the addition of a sticker-spreading agent such as Agral to prevent the fungicide from being washed off the leaf surfaces in wet weather or after watering, however it is not advised when using Funginex.

Soft crown rot (bacterial soft rot) This insidious, devastating disease is caused by the *Erwinia carotovora* bacterium, and it attacks the leaf bases and rhizome, causing the whole plant to disintegrate and collapse. The disease occurs mostly during hot, excessively wet conditions and enters the plant tissue either through stomata or physical damage such as is caused by secateurs, snails, nematodes and caterpillars, and may then be spread from one plant to another by insects such as house flies (Laing 2000). Seemingly healthy plants suddenly begin wilting and fall over, and the disintegration process is accompanied by a strong rotting stench, to which the houseflies are attracted. It is especially prevalent in the summer rainfall parts of southern Africa and usually occurs in poorly drained soil which becomes over-saturated as a result of over-watering or excessive rainfall. The disease rapidly spreads to neighbouring plants and badly affected plants must be destroyed as soon as possible. Plants can often be saved if symptoms are noticed early enough, and if the rhizome has not been too badly affected. Remove the rotting portions by scraping them away with an old scrubbing brush, then wash the affected area thoroughly with water. Apply a liberal coating of captab (e.g. Kaptan) to the affected area, or soak the affected parts in Dithane M45 for half an hour, allow to dry for 30 minutes, then replant in slightly damp, coarse sterilized river sand and place in a shaded, protected spot. New roots should form

Below: Rust attack on leaves of *Clivia miniata*

within a month, and plants should be allowed to fully recover in this medium for about six months, before being planted out into the garden or into permanent pots.

Virus All clivias are highly susceptible to viral disease, the major cause of infection being mealy bug and snail damage and to a lesser extent, damage by aphids, thrips and slugs. Symptoms of viral disease are seen as short or long, dark and light green streaks across the upper surface of the leaf blade and the scape, and may also cause colour-breaking in flowers. Badly affected plants should be destroyed to prevent infection of healthy plants, but if this drastic measure is not desired, affected plants must be kept well isolated and treated preventatively against mealy bugs, snails, aphids, thrips and slugs. If a particularly special plant has become infected, it is best to start again from seed by transferring pollen from a healthy plant onto the infected one, the seeds of which should be free of virus. Bear in mind that viral disease can also be spread by humans on gardening equipment like secateurs, by transferring infected sap from infected to healthy plants. As with other members of the Amaryllidaceae susceptible to viral infection (like *Brunsvigia, Crinum* and *Cyrtanthus*) the symptoms are not immediately visible after infection, often taking up to a year or more to manifest.

Symptoms of viral infection appear as mottled streaks in *Clivia miniata* leaves

REFERENCES AND FURTHER READING

Abel, C. and Abel, J. 2003. Some observations on *Clivia caulescens*. *Clivia* 5: 66–67.

Abel, C. and Abel, J. 2004. Observations on Akebono. *Clivia* 6: 54–56.

Annecke, D.P. and Moran, V.C. 1982. *Insects and mites of cultivated plants in South Africa*. Butterworths, Durban.

Anonymous, 1888. Wien. Illustr. Gartenzeit: 275.

Baker, J.G. 1896. *Clivia*. In: Thiselton-Dyer, W.T. (ed.), *Flora Capensis* 6: 228–229. Reeve & Co., London.

Batten, A. & H. Bokelman, 1966. *Clivia nobilis & C. miniata. Wild Flowers of the Eastern Cape Province*. Books of Africa, Cape Town.

Batten, A. 1986. *Clivia miniata. Flowers of Southern Africa*. Frandsen Publishers, Sandton.

Bayer, A. 1979. *Clivia miniata* var. *flava*. *Flower Paintings of Katharine Saunders*. The Tongaat Group Limited, Maidstone, KwaZulu-Natal.

Behr, E. 1987. *The Last Emperor*. Futura Publications, London.

Blackbeard, G.I. 1939. *Clivia* breeding. *Herbertia* 6: 190–193.

Brown, I. 2003. Promoting early flowering of *Clivia miniata* seedlings. *Clivia* 5: 69–71.

Brown, N.A. and Prosch, D. 2000. Storage and germination of *Clivia miniata* seeds. *Clivia* 2: 59–62.

Burchell, W.J. 1853. *Travels in the interior of southern Africa*. The Batchworth Press, London (reprinted from the original edition of 1822–1824).

Calitz, C. 2003. Visits to special *Clivia* places after the international *Clivia* conference. *Clivia* 5: 57–60.

Christie, M. 2006. Cameron Peach: A brief history. *Clivia News* 15 (2): 7–8.

Chubb, S. 1996. *Clivia* conservation. *Clivia Club Newsletter* 5: 8–9.

Chubb, S. 2002. Plant diseases: Nematodes. *Clivia* 4: 67–68.

Chubb, S. 2004. John Winter: Citation for honorary membership of the Clivia Society. *Clivia Society Newsletter* 13 (2): 10–12.

Chubb, S. 2005. *Clivia miniata*: Colour mutations and their breeding. *Clivia* 7: 75–77.

Chubb, S. 2006. A practical approach to colour breeding in *Clivia miniata*. *Clivia* 8: 52–55.

Chubb, T. and Chubb, S. 2001. Natal peaches. *Clivia* 3: 61–64.

Coates, I. 2004. *Clivia*: food and drink. *Clivia* 6: 52–53.

Conrad, F. and Reeves, G. 2002. Molecular systematics of the genus *Clivia*. *Clivia* 4: 20–23.

Cowley, W. 2006a. *Clivia nobilis* 'Pearl of the Cape'. *Clivia News* 15 (1): 7.

Cowley, W. 2006b. Pale forms of *Clivia nobilis*. *Clivia News* 15 (1): 10–12.

Crouch, N.R., Ndlovu, E., Mulholland, D.A. and Pohl, T. 2003. The genus *Clivia* in ethnomedicine: usage, bioactivity and phytochemistry. *Clivia* 5: 15–22.

De Coster, P. 1999. History of the *Clivia* in Belgium. *Clivia* 1: 31–32.

De Coster, P. 1999. Selection and commercial production of *Clivia* in Europe. *Clivia* 1: 32–39.

De Coster, P. 2002. The ideal pH and nutrition for *Clivia*. *Clivia* 4: 37.

De Smedt, V., van Huylenbroeck, J.M. and Debergh, P.C. 1996. Influence of temperature and supplementary lighting on growth and flower initiation of *Clivia miniata* Regel. *Scientia Horticulturae* 65: 65–72.

Desmond, R. 1994. *Dictionary of British & Irish botanists and horticulturists*. Taylor & Francis, London.

Dixon, R. 2005a. Odd umbels. *Clivia* 7: 29.

Dixon, R. 2005b. *Clivia robusta* 'Maxima'. *Clivia* 7: 62.

Dixon, R. 2005c. Have genes, will travel: On the trail of 'Vico Yellow'. *Clivia* 7: 78–81.

Dixon, R. 2006a. Types of variegation in *Clivia*. *Clivia Yearbook* 8: 73–76.

Dixon, R. 2006b. *Clivia*. South African Post Office. http://www.sapo.co.za

Dower, M. and Robbertse, H. 2003. Inflorescence bud abortion and the importance of growth modules when dividing and feeding *Clivia miniata*. *Clivia* 5: 105–107.

Duncan, G.D. 1985. Notes on the genus *Clivia* Lindl. with particular reference to *C. miniata* Regel var. *citrina* Watson. *Veld & Flora* 71(3): 84–85.

Duncan, G.D. 1991. Clivias and their cultivation. *Parks & Grounds* 59: 21–22.

Duncan, G.D. 1992a. Clivias in Kyushu and welwitschias in Kyoto. *Veld & Flora* 78(3): 76–77.

Duncan, G.D. 1992b. Notes on the genus *Clivia* Lindl. with particular reference to *C. miniata* Regel var. *citrina* Watson. *Herbertia* 48: 26–29.

Duncan, G.D. 1996. *Growing South African bulbous plants: A popular guide*. National Botanical Institute, Cape Town.

Duncan, G.D. 1999. *Grow Clivias*. Kirstenbosch Gardening Series. National Botanical Institute, Cape Town.

Duncan, G.D. 2000a. *Grow Bulbs*. Kirstenbosch Gardening Series. National Botanical Institute, Cape Town.

Duncan, G.D. 2000b. *Cryptostephanus vansonii*: As curious amaryllid from Zimbabwe. *Veld & Flora* 88 (1): 18–19.

Duncan, G.D. 2004a. *Amaryllis* magic. *Veld & Flora* 90(4): 142–147.

Duncan, G.D. 2004b. Growing the miracle *Clivia*. *Veld & Flora* 90(4): 148–149.

Duncan, G.D. 2005. Character variation and a cladistic analysis of the genus *Lachenalia* Jacq. f. ex Murray (Hyacinthaceae: Massonieae). M.Sc. thesis, University of KwaZulu-Natal, Pietermaritzburg.

Duncan, G.D. 2006. Bulbous wealth at the Cape. *The Alpine Gardener* 74 (3): 296–315.

Du Plessis, N.M. & Duncan, G.D. *Clivia*. *Bulbous plants of southern Africa*: 96–97. Tafelberg, Cape Town.

Dyer, R.A. 1943. *Clivia caulescens*. The *Flowering Plants of South Africa* 23 t. 891.

Dyer, R.A. 1976. *The genera of southern African flowering plants*. 2. Gymnosperms and monocotyledons, *Clivia*: 950. Botanical Research Institute, Pretoria.

Faber, G. 2007. *Clivia* and the implications of the new RSA Legislation. *Clivia News* 16(2): 10–13.

Finnie, J.F. 1998. *In vitro* culture of *Clivia miniata*. *Clivia* 1: 7–10.

Fisher, R.C. 2004. Pollination by moths? *Clivia Society Newsletter* 13 (2): 18–19.

Fisher, R.C. 2006. The seeing hand: *Clivia* depicted. *Clivia* 8: 100–107.

Forssman, C. 1948. The clivias at Scott's Farm, Grahamstown. *Herbertia* 15: 59–63.

Gerber, J. 2003. Insect control on *Clivia*. *Clivia* 5: 101–103.

Giddy, C. 1982. The noble *Clivia*. *Natal Gardener*: 10.

Glover, W.J. 1985. Considerations in *Clivia*. *Herbertia* 41: 30–31.

Gouws, J.B. 1949. The karyology of some South African Amaryllidaceae. *Plant Life* 5: 54–81.

Grebe, H. 2004. In search of a 'Light of Buddha'. *Clivia* 6: 74–80.

Grebe, H. 2005. In search of *Clivia mirabilis*: Some observations from the wild. *Clivia* 7: 38–44.

Grebe, H. 2006. *Clivia mirabilis* in the Western Cape. *Clivia* 8: 19 22.

Griffiths, M. (ed.) 1992. *Clivia*. *Royal Horticultural Society Dictionary of Gardening* 1: 657–658. Macmillan, London.

Gouws, J.B. 1949. Karyology of some South African Amaryllidaceae. *Plant Life* 5: 48.

Grobler, A. and van der Merwe, L. 2002. Vegetative (asexual) propagation of *Clivia*. *Clivia* 4: 24–26.

Groenland, J. 1859. *Himantophyllum miniatum*. *Revue Horticole* 125: tt. 29–30.

Gunn, M. and Codd, L.E. 1981. *Botanical exploration of southern Africa*: 371. A.A. Balkema, Rotterdam.

Hammett, K.R.M. 2001. Swamp *Clivia*. *Clivia* 3: 69–72.

Hammett, K.R.M. 2002. *Clivia*. *Bulbs* 4: 20–28.

Hammett, K.R.M. 2005. *Clivia robusta*: The 'Swamp Clivia'. *Clivia* 7: 56–61.

Hammett, K.R.M. 2006. Pigment surprise. *Clivia Yearbook* 8: 39–49.

Herbert, W. 1837. *Amaryllidaceae*. James Ridgeway & Sons, London.

Hooker, W.J. 1828. *Imatophyllum aitoni*: Handsome-flowered *Imatophyllum*. *Curtis's Botanical Magazine* 55: t. 2856.

Hooker, W.J. 1854. *Imantophyllum ? miniatum*: Brick-coloured *Imantophyllum*. *Curtis's Botanical Magazine* 80: t. 4783.

Hooker, W.J. 1856. *Clivia gardeni*. *Curtis's Botanical Magazine* series 3, 12: t. 4895.

Honiball, C. 2001. Manipulation of flowering period and shoot multiplication in *Clivia miniata* Regel. *Clivia* 3: 30–56.

Hutchings, A., Scott, A.H., Lewis, G. and Cunningham, A.B. 1996. *Zulu medicinal plants*. University of Natal Press, Pietermaritzburg.

Hutchinson, J. 1946. *A botanist in southern Africa. The travels of William J. Burchell*: 625–633. P.R. Gawthorn, London.

IUCN 2006. *IUCN Red List categories*. IUCN Species Survival Commission, Gland, Switzerland.

Jeans, M. 2004. *Clivia miniata*: in favour of seed grown strains. *Clivia* 6: 39–42.

Kirk, J.T. and Tilney-Bassett, R.A. 1978. *The plastids. Their chemistry, structure, growth and inheritance*. Elsevier and North-Holland Biomedical Press, Amsterdam.

Koopowitz, H. 2000. *Clivia* names. *Clivia* 2: 30–33.

Koopowitz, H. 2002. *Clivias*. Timber Press, Oregon.

Koopowitz, H., Griesbach, R. and Comstock, J. 2003. Colour pigments in *Clivia*. *Clivia* 5: 23–31.

Laing, M. 2000. Bacterial soft rot of clivias. *Clivia* 2: 64–66.

Leighton, F.M. 1939. The history of botanical exploration for amaryllids in South Africa. *Herbertia* 6: 15–24.

Lindley, J. 1828. *Clivia nobilis*, scarlet clivia. *Edwards's Botanical Register* 14: t. 1182.

Lindley, J. 1854. *Vallota ? miniata*. The *Gardener's Chronicle* 8: 119.

Lötter, C. 1998. A comprehensive discussion of the cultivation of clivias. *Clivia* 1: 24–28.

Lötter, W.J. 1998a. *Clivia* mutations and colour variations. *Clivia* 1: 63–72.

Lötter, W.J. 1998b. Breeding behaviour of 'Natal Yellow'. *Clivia Club Newsletter* 7(1): 12–14.

Lötter, W.J. 2000. Advanced hybridizing of clivias. *Clivia* 2: 34–41.

Lötter, W.J. 2003. *Clivia* mutations and modifications. *Clivia* 5: 87–89.

MacDermott, J. 2004. Fungicide use in clivias. *Cape News* 5: 9–11.

Malan, C. 2000a. The Eastern Cape: of its people and plants. *Clivia* 2: 15–17.

Malan, C. 2000b. Gladys Blackbeard: *Clivia* pioneer. *Clivia* 2: 42.

Manning, J. 2005. Pollination in *Clivia*. *Clivia* 7: 17–22.

Marriott, H. 2006. Nakamura's contribution to *Clivia* breeding. *Clivia* 8: 6–18.

May, R. 2001. *Brithys crini* (Fab.) (Lepidoptera: Noctuidae) in Australia: from the bush to suburbia. *Clivia Club Newsletter* 10 (3): 8–9.

McCracken, D.P. and McCracken, P.A. 1990. *Natal, the garden colony*. Frandsen Publishers, Johannesburg.

McNeil, A. 1998. Gordon McNeil. *Clivia Club Newsletter* 7(2): 19–20.

McNeil, P.G. 1963. *Cryptostephanus vansonii*. *Plant Life* 19: 54.

McNeil, P.G. 1985. Hybridising clivias. *Herbertia* 41: 24–29.

Morris, W. 1990. A true breeding strain of yellow *Clivia*. *Herbertia* 46(2): 95–96.

Morris, W. 1993. Variegation. *Clivia Club Newsletter* 2(3): 4–5.

Morris, W. 1994. A hybrid by any other name. *Clivia Club Newsletter* 3(7): 1–2.

Morris, W. 1997. 1996 Yellow *Clivia* report. *Herbertia* 52: 66–68.

Morris, W. 1998a. A theory on the genetics of yellow *Clivia*. *Clivia Club Newsletter* 7(2): 11–14.

Morris, W. 1998b. The Cowlishaw clivias. *Herbertia* 53: 164–167.

Morris, W. 1999. Classification of the known varieties of yellow clivia. *Clivia Club Newsletter* 8(1): 10–13.

Clivia robusta var. *robusta* from the Eastern Cape

Morris, W. 2002. Peach and pastel *Clivia* and their origins. *Clivia* 4: 27–30.

Müller-Doblies, D. 1980. Notes on the inflorescence of *Agapanthus*. *Plantlife* 36: 72–76.

Murray, B.G., Ran, Y., De Lange, P.J., Hammett, K.R.W., Truter, J.T. and Swanevelder, Z.H., 2004. A new species of *Clivia* (Amaryllidaceae) endemic to the Pondoland Centre of Endemism, South Africa. *Botanical Journal of the Linnean Society* 146: 369–374.

Nakamura, Y. 2000. *Clivia* in China. *Clivia* 2: 69–70.

Nel, S. 2006a. *Clivia* stamps around the globe. *Clivia News* 15 (1): 7–8.

Nel, S. 2006b. Reports of additional clivias on stamps. *Clivia News* 15 (4): 23.

Ogasawara, R. 1997. *Clivia* (in Japanese). NHK Broadcasting Publication, Tokyo.

Ogasawara, S. 2008. *Clivia* (in Japanese). NHK Broadcasting Publication, Tokyo.

Obermeyer, A.A. 1970. *Clivia gardenii*. *The Flowering Plants of Africa* 42: t.1641.

Pooley, E. 1998. A field guide to wildflowers of KwaZulu-Natal and the Eastern Region: 36–39. Natal Flora Publications Trust, Durban.

Phillips, E.P. 1921. *Clivia miniata*. *The Flowering Plants of South Africa* 1: t.13.

Phillips, E.P. 1931. *Clivia miniata* var. *flava*. *The Flowering Plants of South Africa* 11: t. 411.

Ran, Y., Murray, B.G. and Hammett, K.R.W. 1998. Research in *Clivia* chromosomes. *Clivia* 1: 48–55.

Ran, Y., Murray, B.G. and Hammett, K.R.W. 1999. Karyotype analysis of the genus *Clivia* by giemsa and fluorochrome banding and *in situ* hybridization. *Euphytica* 106: 139–147.

Ran, Y., Hammett, K.R.W. and Murray, B.G. 2001. Hybrid identification in *Clivia* (Amaryllidaceae) using chromosome banding and genomic *in situ* hybridisation. *Annals of Botany* 87: 457–462.

Regel, E.A. 1864. *Clivia miniata* (Lindl.) Regel. *Gartenflora* 14: 131, t. 234.

Robbertse, H. 2001a. *Clivia* terminology. *Clivia* 3: 11–12.

Robbertse, H. 2001b. The *Clivia* pistil: structure and function. *Clivia* 3: 16–19.

Robbertse, H. 2002. The nutrition system of *Clivia*. *Clivia* 4: 31–35.

Robbertse, H. and van der Merwe, C. 2003. Leaf formation and its relation to flowering in *Clivia miniata*. *Clivia* 5: 42–46.

Robbertse, H. and Zwanevelder, Z.H. 2001. *Clivia* pollen: function and structure. *Clivia* 3: 13–15.

Robbertse, H. and Pienaar, M. 2003. Notes on pollination and pollen tube growth in *Clivia miniata*. *Clivia* 5: 81–82.

Roemer, J.A. and Schultes, J.H. 1830. *Systema vegetabilium* 7: 892. Cotta, Stuttgart.

Rourke, J.P. 2002a. *Clivia mirabilis* (Amaryllidaceae: Haemantheae) a new species from Northern Cape, South Africa. *Bothalia* 32(1): 1–7.

Rourke, J.P. 2002b. The miraculous *Clivia*, an astonishing new species from the arid Northern Cape. *Clivia* 4: 5–12.

Rourke, J.P. 2003a. Natural interspecific hybrids in *Clivia*. *Clivia* 5: 78–80.

Rourke, J.P. 2003b. Secondary pseudo-umbels in *Clivia mirabilis* inflorescences. *Clivia* 5: 91.

Sasaki, S. 2001. *Clivia* in Japan. *Clivia* 3: 24–29.

Sasaki, S. 2003. Variegated *Clivia* in Japan. *Clivia* 5: 52–56.

Sasaki, S. 2004. Multitepal breeding. *Clivia* 6: 57–60.

Sasaki, S. 2005. New developments in *Clivia* breeding in Japan. *Clivia* 7: 86–90.

Shiang, S. and Song-liang, 1999. '*Variegated Clivia*' (In Chinese). Jilin Science Technology Publications, Changchun.

Smith, C.A. 1966. Common names of South African plants. *Clivia* Lindl.: 543. Dept. Agricultural Technical Services, Pretoria.

Smith, K.R. 2003a. *Clivia* in Australia. *Clivia* 5: 11–14.

Smith, K.R. 2003b. The caulescens tour. *Clivia* 5: 61–62.

Smith, K.R. 2003c. The name game. *Clivia* 5: 108–111.

Smith, K.R. 2004. Maternal inheritance: more observations about variegated *Clivia*. *Clivia* 6: 30–32.

Smith, K.R. 2006. *Clivia* Cultivar checklist and registration. *Clivia News* 15 (2): 12.

Smithers, P. 1995. The origins of *Clivia miniata* 'Vico Yellow' and 'Vico Gold'. *Herbertia* 50: 9–12.

Smithers, P. 2000. Adventures of a gardener. *Clivia Yearbook* 2: 13–14.

Snijman, D.A. 2002a. *Clivia*. In: Leistner, O.A., ed., Seed plants of southern Africa: families and genera. *Strelitzia* 10: 572–573.

Snijman, D.A. 2002b. A remarkable new discovery in *Clivia*. *Herbertia* 57: 35–40.

Snijman, D.A. 2003. Fire and the distribution of *Clivia* in southern Africa. *Clivia* 5: 98–100.

Snijman, D.A. and Archer, R.H. 2003. *Clivia*. In: Germishuizen, G. and Meyer, N.L. Plants of southern Africa: an annotated checklist. *Strelitzia* 14: 958. National Botanical Institute, Pretoria.

Snijman, D.A. and Victor, J.E. 2003. Threatened Amaryllidaceae of South Africa. *Herbertia* 58: 91–108.

Spies, J. 2004. Breeding broad leaf *Clivia*: theoretical genetic considerations. *Clivia* 6: 33–38.

Spies, J. 2005a. Genetic variation in *Clivia*. *Clivia* 7: 6–11.

Spies, J. 2005b. What's in a name? *Clivia* 7: 32–34.

Spies, J. 2006. Genetic aspects of *Clivia* breeding. *Clivia* 8: 31–38.

Ströh, H. 2002. Pests and diseases affecting *Clivia* in South Africa. *Clivia* 4: 69–80.

Swanevelder, Z.H. 2003. Diversity and population structure of *Clivia miniata* Lindl. (Amaryllidaceae): evidence from molecular genetics and ecology. M.Sc. thesis, University of Pretoria, Pretoria.

Swanevelder, Z.H. 2004. Horticultural significance of *Clivia miniata*. *Clivia* 6: 8–16.

Swanevelder, Z.H., Van Wyk, A.E. and Truter, J.T. 2005. A new variety in the genus *Clivia*. *Bothalia* 35 (1): 67–68.

Swanevelder, Z.H., Forbes-Hardinge, A., Truter, J.T. and van Wyk, A.E. 2006. A new variety of *Clivia robusta*. *Bothalia* 36 (1): 66–68.

Swanevelder, Z.H., Truter, J.T. and van Wyk, A.E. 2006. A natural hybrid in the genus *Clivia*. *Bothalia* 36 (1): 77–80.

Swart, W.J. 2004. Diagnosing plant disease: what the grower needs to know. *Clivia* 6: 61–66.

Swart, W.J. 2005. Fungal pathogens associated with *Clivia*. *Clivia* 7: 91–96.

Synge, P.M. ed.1950. Plants to which awards have been made in 1950: *Clivia* x *kewensis* 'Bodnant Yellow'. *Journal of the Royal Horticultural Society* 75(8): 334.

Tarr, B.B. 2000. The great yellow clivia jigsaw puzzle. *Clivia Yearbook* 2: 10–12.

Tarr, B.B. 2005. *Clivia gardenii*: autumn's delight. *Clivia* 7: 63–66.

Thurston, V.A. 1998. *The Clivia*. Privately printed, Tongaat, KwaZulu-Natal.

Traub, H.P. 1976. Amaryllid notes. *Plant Life* 32: 57.

Truter, J.H., Swanevelder, Z.H. and Pearton, T.N. 2006. *Clivia* x *nimbicola*: a stunning beauty from the Bearded Man. *Clivia* 8: 23–27.

Van der Linde, J. 2001. The great Chinese *Clivia* 'bubble'. *Clivia* 3: 20–23.

Van der Linde, J. 2003a. Some early names associated with *Clivia* (3): William John Burchell. *Clivia Society Newsletter* 12 (1): 18–19.

Van der Linde, J. 2003b. Was the plant named *Clivia nobilis* in 1828 surreptitiously obtained from Kew? *Clivia* 5: 92–95.

Van der Linde, J. 2004a. The story behind the naming of *Clivia miniata*. *Bulletin of the Indigenous Bulb Association of South Africa* 53: 14–15.

Van der Linde, J. 2004b. Design your own *Clivia*. *Clivia* 6: 17–22.

Van der Linde, J. 2004c. More on mirabilis. *Clivia* 6: 23.

Van der Linde, J. 2004d. Personality profile – Sir Peter Smithers. *Clivia Society Newsletter* 13 (1): 24–25.

Van der Linde, J. 2005. Why do *Clivia* flower when they do? *Clivia* 7: 27–28.

Van der Linde, J. 2006. The importance of the flower pedicels in selective breeding of variegated *Clivia*. *Clivia News* 15 (1): 14–15.

Van der Linde, J. 2007. Two classic clivias at Kirstenbosch and some of their progeny. *Clivia* 9: 59–63

Van der Merwe, L., Robbertse, H. and de Kock, B. (eds) 2005. *Cultivation of Clivias*. Clivia Society, Pretoria.

Van Houtte, L. 1869. *Imatophyllum* (Hybr.) Cyrtanthiflorum. *Flore des Serres et Jardins de l'Europe* (Series II) 87 t. 1877. Louis Van Houtte, Ghent.

Van Huylenbroeck, J.M. 1998. *Clivia miniata* Regel: control of plant development and flowering. *Clivia* 1: 13–20.

Van Rensburg, D. 2002. The role of light (or radiation) on the growing of plants. *Clivia* 4: 60–66.

Van Voorst, A. 2003. An introduction to polyploidy in *Clivia* breeding. *Clivia* 5: 33–39.

Van Voorst, A. 2004. Polyploidy in *Clivia*: a layman's guide. *Clivia* 6: 43–46.

Van Voorst, A. 2006. Ploidy research in *Clivia*: an update. *Clivia* 8: 56–63.

Van Wyk, B-E., B. Van Oudtshoorn and N. Gericke, 1997. *Medicinal plants of South Africa*: 88–89. Briza, Pretoria.

Van Wyk, A.E. and Smith, G.F. 2001. *Regions of floristic endemism in southern Africa. A review with emphasis on succulents*. Umdaus, Pretoria.

Verdoorn, I.C. 1943. *Cryptostephanus vansonii*. *The Flowering Plants of South Africa* 23: t. 885.

Victor, J.E. and Keith, M. 2004. The Orange List: a safety net for biodiversity in South Africa. *South African Journal of Science* 100: 139–141.

Vildomat, F., Bastida, J., Codina, C., Nair, J. and Campbell, W. 1997. Alkaloids of the South African Amaryllidaceae. *Recent Research and Development in Phytochemistry* 1: 131–165.

Vissers, M. 2002. The role played by the various elements of nutrition in the life, growth and health of plants. *Clivia* 4: 44–59.

Vorster, P. and Smith, C. 1994. *Clivia nobilis*. *Flowering Plants of Africa* 53: 70–74, t. 2094.

Wager, V.A. 1977. *Agapanthus* leaf die-back. *Plant Pests and Diseases*: 99. Jonathan Ball, Johannesburg.

Watson, W. 1899a. *Cliveia miniata* var. *citrina*. *The Gardener's Chronicle* (3rd series) 25: 228.

Watson, W. 1899b. *Clivia miniata* var. *citrina*. *The Garden* 56: 338–339, t. 1246.

Williams, V. 2005. *Clivia* under threat. *Clivia* 7: 12–16.

Winter, J.H. 2000. The natural distribution and ecology of *Clivia*. *Clivia* 2: 5–9.

Winter, J.H. 2002. Growing and propagating *Clivia mirabilis*. *Clivia* 4: 13–14.

Winter, J.H. 2003a. Collecting *Clivia* in their natural habitat. *Clivia* 5: 5–10.

Winter, J.H. 2003b. Persistent mealybugs. *Clivia* 5: 103.

Winter, J.H. 2005a. *Clivia caulescens*. Plantzafrica.com

Winter, J.H. 2005b. *Clivia mirabilis*. Plantzafrica.com

Zonneveld, B.J.M. 2002. The systematic value of nuclear DNA content in *Clivia*. *Herbertia* 57: 41–48.

Zonneveld, B.J.M. 2005. Nuclear DNA content in *Clivia*. *Clivia* 7: 35–37.

Zonneveld, B.J.M. 2006. Variegation in *Clivia*. *Clivia* 8: 66–72.

Zonneveld, B.J.M. and Duncan, G.D. 2003. Taxonomic implications of genome size and pollen colour and vitality for species of *Agapanthus* L'Héritier (Agapanthaceae). *Plant Systematics and Evolution* 241: 115–123.

Zonneveld, B.J.M. and Duncan, G.D. 2006. Genome size for the species of *Nerine* Herb. (Amaryllidaceae) and its evident correlation with growth cycle, leaf width and other morphological characters. *Plant Systematics and Evolution* 257: 251–260.

Clivia miniata Kirstenbosch Supreme strain

GLOSSARY

acaulescent: without an aerial stem

actinomorphic: regular, flowers capable of being bisected into similar halves in more than one vertical plane

acute: sharply pointed, narrowing gradually

albinism: an inherited lethal condition resulting in a complete lack of chlorophyll

alkaloids: compounds with medicinal and toxic properties, found in a wide variety of plants, e.g. isoquinoline alkaloids in *Clivia*

allopatric: two or more taxa having different distribution ranges

anther: the part of the stamen that contains pollen

anthesis: period during which the flower is fully open

apex: the terminal end

apiculate: ending abruptly in a short point or apiculum

berry: fruit with soft flesh surrounding one or more seeds

bracteole: secondary bract sheathing a flower in an inflorescence, itself enclosed within a primary bract e.g. spathe valve

buttress roots: stabilizing, adventitious roots produced in waterlogged soil, e.g. in certain forms of *Clivia robusta*

canaliculate: channelled

caulescent: an aerial stem bearing leaves

chloroplast: an organelle found in plant cells and certain algae that conducts photosynthesis

chlorosis: loss of chlorophyll and consequently loss of green leaf colouration due to mineral deficiency, insufficient or excessive light, or disease

clone: individuals that are genetically identical, derived from one individual by vegetative propagation

cotyledon: the primary (seedling) leaf

cross-pollination: when the stigma of a flower on one plant is brushed with pollen from the flower of a different plant

cultivar: a cultivated variety; a plant raised or selected in cultivation that retains distinct, uniform characteristics when propagated; all plants carrying a cultivar name must be genetically identical

cymbiform: boat-shaped

declinate: bent downwards or forwards

distichous: leaves arranged in two opposite rows, one leaf above the other

dorsifixed: refers to the anthers that are attached to the filaments at their backs

drooping: with the apex directed towards the horizon

endosperm: tissue that surrounds and nourishes the embryo in seeds

epiphyte: a plant that grows on another plant but does not derive nourishment from it

etymology: derivation and original meaning of words

exserted: protruding beyond a surrounding organ, e.g. stamens exserted beyond the perianth

F1 hybrid: first-generation offspring that is vigorous and uniform, obtained by crossing two distinct, pure-bred lines

F2 hybrid: offspring obtained by self-pollination within a population of F1 hybrids, which do not come true

filament: the stalk of a stamen that bears the anther

filiform: thread-like

glabrous: smooth and hairless

glaucous: having a grey, blue-grey, blue-green or white bloom

gross feeders: plants that respond favourably to heavy feeding

half-hardy: able to withstand temperatures down to 0°C

hardiness: ability of a cultivated plant to withstand adverse conditions, e.g. cold tolerance

inflorescence: the arrangement of the flowers

infructescence: the arrangement of fruits on a plant

intergeneric hybrid: result of crossing plants of two distinct, usually closely related genera

interspecific hybrid: result of crossing two species within the same genus

keel: a ridge, like the keel of a boat

lanceolate: lance-shaped, broadest in the lower half, narrowing towards the apex

lateral: located on the side of an organ, e.g. lateral inflorescence produced on pseudostem of *Clivia*

linear: long and narrow, with parallel or almost parallel margins

lithophyte: a plant that usually grows on or among rocks, e.g. *Clivia caulescens*

locule: a compartment of the ovary containing ovules

median: in the middle

meristem: the growing region of a plant, in which cells divide to produce new cells

microclimate: local environmental conditions in a specific and limited area, which may differ greatly from general conditions in the area

mutation: a natural or artificially induced genetic change, e.g. a *Clivia* seedling with variegated leaves

node: point on a stem at which leaves and shoots arise

oblanceolate: inversely lanceolate, with the narrowest portion at the base and the broadest portion nearest the apex

offset: small plant that arises from mother plant by vegetative means

ovoid: egg-shaped

ovule: structure in the ovary from which seed develops after fertilization

outgroup: one or more taxa hypothesized to be less closely related to all the taxa in the ingroup, used to improve the polarity of characters in cladistic analysis

papyraceous: papery

pedicel: the stalk of an individual flower

peduncle: the stalk of an inflorescence

pendent: hanging downwards, due to the weight of the flower or fruit

perianth: the floral envelope, all the tepals considered together

pericarp: protective outer tissue that develops around the seeds from the ovary wall of the flower

picotee: a flower with a narrow band of contrasting colour around the margin of each tepal

pollination: the transfer of pollen from the anthers to the stigma of the same or a different flower

pseudostem: a false stem that looks like a stem but is composed of tightly packed leaf sheaths

radicle: the rudimentary root of a germinating seed

re-curved: arched backwards

rhizome: a specialized vertical stem growing at or below ground level

scape: a synonym of peduncle

selection: a distinct form of a plant that is selected for its superior qualities, then propagated (usually vegetatively) and given a cultivar name

self-sterile: a plant that requires pollen from a different plant (but not of the same clone) to fertilize its flowers

serrated: toothed, with regular, pointed teeth, e.g. *Clivia nobilis*

sheath: a tubular structure around the base of a plant, e.g. a leaf base around the stem in *Clivia*

spathe bract: one or more primary bracts enclosing the inflorescence

spathulate: spoon-shaped, rounded above and narrow below

species: a taxon comprising similar individuals that breed true to type in the wild

stamen: the male part of a flower, comprising an anther and a filament

stolon: horizontal stem produced just below soil level from the base of the mother plant, resulting in a new plant at the tip

strain: the stock of a particular form, selected with special attention to some important character and breeding true to it, e.g. the short, broad-leafed *C. miniata* hybrid 'Daruma' strain

sympatric: two or more taxa having overlapping distribution ranges

taxon: a category at any level used to classify organisms, e.g. variety, subspecies, species

tepal: a collective term for a petal and sepal of a flower where the calyx and corolla are not clearly distinguished

terrestrial: a plant that grows on or in the ground

tissue culture: growth of plant cells in an artificial medium under sterile laboratory conditions

tricuspidate: refers to the stigma that is divided into three cusps, each tipped with an abrupt, rigid point

umbel: an inflorescence in which all the pedicels arise from a common point at the top of the peduncle, and are of similar length

undulate: refers to the wavy margins of leaves or flowers

variegation: leaves and stems marked with irregular lines of different colours

variety: a naturally occurring variant of a species, showing a recognizable character e.g. yellow pigmentation, taxonomically ranked between subspecies and *forma*

vector: distributor

velamen: water- and nutrient-retaining corky outer layer of roots that prevents desiccation of root tissue

versatile: moving freely around the point of attachment to the filament

xerophytic: a plant adapted to survive arid conditions, e.g. *Clivia mirabilis*

zygomorphic: irregular, flowers capable of being bisected into similar halves in one vertical plane only

INDEX

Page numbers in **bold** denote illustrations

A
advanced hybrids 101, 103,
Afrika, Johannes 13
Agapanthus 24, 27, 28, 31
　africanus 48
　fungus 64, **164**
　praecox 30, 138
Aiton, William Townsend 3
Akebono variegation 124, 125
Albany Centre of Endemism 82
Alstroemeria 17
Amaryllidaceae 84
Amaryllidaceae 21, 168
Amaryllis 40
amaryllis caterpillar see lily borer moth
Ammocharis coranica 160
Andrews, James 4
aphids 158
Apis mellifera 42
Argentine ants 161
artificial interspecific hybrids see interspecific hybrids
artificial intraspecific hybrids see intraspecific hybrids

B
Backhouse Nurseries, York 5
Backhouse, James 5, 11
Backhouse, Thomas 5, 11
bacterial soft rot see soft crown rot
Baker, J G 11
Barberton Centre of Endemism 104.
bee pollination see honey bee pollination
Belgian Hybrids 7, 112, 117
bird pollination syndrome 41
black sunbird 40
Blackbeard, Gladys L 16
blackfly see aphids
Botanical Journal of the Linnean Society 13
Bothalia 13, 15, 76, 91, 96, 103
Bowden, W H 8
Bowie, James 3, 4
Brithys pancratii see lily borer moth
Brunsvigia 168
Burchell, William John 1, 3,

bush lily see *Clivia nobilis*
butterfly pollination 41, 42, 103

C
Cape Clivia Club iii, 18
caper spurge see *Euphorbia lathyris*
Chiazzari,W L 13
Chiliza, Sibonelo 77
chlorosis 164, **165**
Christie, Michael 112
Chubb, Sean 46, 105, 121, 122
Clive, Lady Charlotte Florentia, Duchess of Northumberland 3, 82,
Clivia (Journal of the Clivia Society) 117
　Lindl. 55
　and *Hippeastrum* 'Green Girl' 16
　hybrid (*C. gardenii* x *C. miniata*) 101
　hybrid (*C. miniata* x *C. mirabilis*) 19
　hybrids see also hybrid clivias
　x *cyrtanthiflora* 15
　x *nimbicola* Z.H.Swanevelder et al. **13**, 15, 21, **23**, 47, 101, **102**, 103, 104
　Caulgard Group (*C. gardenii* x *C. caulescens*) 107
　Cyrtanthiflora Group (*C. nobilis* x *C. miniata*) **15**, 101, **106, 107**, 108, **133**
　[Group unknown] 'Crayon' **112**
　[Group unknown] 'Soko Jiro' 112, **113**
　Minicyrt Group (*C.* [Cyrtanthiflora Group] x *C. miniata*) 101, 107
　Minigard Group (*C. gardenii* x *C. miniata*) 107
　[Minigard Group] 'Journey' **109**
　Minilescent Group (*C. miniata* x *C. caulescens*) 107
　[Minilescent Group] 'Mandala' **cover, 109, 110**
　Minirabilescent Group (*C. mirabilis* x *C. miniata*) **ii**, 107, **111**
　Nobilescent Group (*C. nobilis* x *C. caulescens*) 107
　Noble Guard Group (*C. gardenii* x *C. nobilis*) 107

　Clivia caulescens R.A.Dyer **back cover, 10**, 11, 13, 15, 18, 21, **22, 26**, 27, 28, 30, **31**, 32, 33, 35, 36, **37**, 38, 40, 42, **43**, 45, 47, 49, 50, 51, 56, **71**, 72, **73**,

74, 103, 104, 109, 123, 135, 136, **137, 138,** 139, 141, 144, 149, 150, 154, 159

Clivia gardenii W.J.Hooker **cover, 6,** 7, 13, 18, 21, 27, 28, 30, **32,** 35-**37, 38,** 40, 45, **46,** 49-51, 54, 56, 91, 96, 103, 104, 109, 134, 136, 137, 138, 139, 141, 144, 149, 150, 154, 166
var. *citrina* Z.H.Swanevelder *et al.* **iii**, 13, 21, **23,** 33, **39,** 47, 91, 94, 96, **97,** 98, **99**
var. *gardenii* **22,** 46, 51, **93,** 94, **95,** 98

Clivia miniata (Lindl.) Regel **iv,** 4, 5, 7, 11, 15-18, 21, 24, 27, 28, **29,** 30, 32, **33, 34,** 35, **36, 37,** 40-42, 44, **45,** 46, 47, **49,** 50, 56, 72, 80, 82, 94, 96, **100,** 101, 103-105, 107-110, 112, 114, 120, 121, 123, **132,** 135, **136,** 137, 138, 139, 141, 144, **146, 147, 149**-152, 154, **157, 159,** 161, **162, 163, 164, 165,** 166, 167, **168**
hybrid cover, 11, 33, 100, 105, 145
(light orange form) 53
(pastel coloured forms) 60
(pastel coloured hybrids) 19, 33, 38, 60
(red form) 55
(yellow form) 7, 8
multitepal 130
pygmies 131
'Admiration' 11
'Akebono-fu' **123, 125, 151**
'Andrew Gibson' **100**
Cameron Peach strain 112, **113**
'Cheryl Apricot' 112
'Chubb's Peach' **62,** 120
'Determination' 122
Daruma strain 114, 150
Daruma strain (yellow form) 114
'Favourite' 11
'Frats' tepal **131**
'Fukurin-fu' **123, 126**
'Genpei-fu' **127**
'Hirao' **114**
'Ito fukurin-fu' **129**
'Kewensis A' 118

'Kewensis B' 118
'Kewensis Cream' 118
'Kirstenbosch Supreme' **i, 20,** 66, **115,** 116
Kirstenbosch Supreme strain 116, **176, 187**
'Light of Buddha' 124, **125**
'Marie Reimers' 11
'Mars' **122**
'Morné Bouquex' 122
'Naka-fu' **127**
'Naude's Peach' 122
'Negishi-fu' **127**
'Optima' 11
Pat's Gold strain **cover, vi,** 66, **116,** 117, **148**
'Shima-fu' **29,** 127, **128,** 141
'Skwebizi Bicolor' **121**
'Smither's Yellow' 119
'So Excited' **121**
'Soft Touch' **122**
'Soko Jiro' **117**
'TK Best Bronze' **117,** 117
'TK Yellow' 118, **119**
'Tora-fu' **129**
'Tora-fu' ('Taihou') **129**
'Vico Gold' 119, 120
'Vico Peach' **120**
'Vico Yellow' 101, **118**-119, **121,** 155
'White Ghost' **120**
var. *citrina* **9,** 11, 13, 21, **23,** 27, **38,** 39, 42, 45, 62, 91, 98, 104, 120, 136, 151, 154, **155, 165**
'Aurea' 121
'Butter Yellow' **65,** 66, 117
'Dwesa Yellow' 64, 66, **67**
'Kirstenbosch Yellow' **35,** 42, 66, **68,** 70, 116, 164
'Mare's Yellow' 66
'Mvuma Yellow' 66
'Natal Yellow' **cover,** 33, 38, 66, **69,** 116, 117, 122, 154
'Ndwedwe Alpha Thurston' 67, **69**
'Ndwedwe Beta Thurston' 67
'Noyce's Sunburst' **70**
var. *flava* 8
var. *miniata* **22, 42,** 45, 51, **57-**64, 104, 151
var. *miniata* 'Appleblossom' **cover,** 60,

61, 103
'Chubb's Peach' 60, **62**
'Mbashe Peach' 60, 62, **63**
'Naude's Peach' 62, 67
'Ndwedwe Gamma Peach' 62

Clivia mirabilis Rourke **cover,** iii, 11, **12,**
   13, 18, 19, 21, **22,** 24, **26, 27, 28, 30,**
   31-33, **35, 36,** 38, 40, 42, **47, 48,** 50-
   52, 56, **75,** 76, **77, 78**-80, 84, 110, **134,**
   135, 139, 141, 142, **143,** 144, 148,
   149, 150, 154, 157, 159

Clivia nobilis Lindl. **1, 2,** 3, 5, 15, 18, 21,
   22, 26-**28,** 29, 30, 32, 33, 35, 36, **38,**
   40, 42, **44,** 45, 48, 49, **50,** 51, 55, 56,
   79, 80, **81,** 82, **83, 85, 86,** 103, 107,
   108, 135, 136, 137, 138, 139, 141,
   144, 149, 150, **159**
   (orange-pink form) 85
   (pinkish-yellow form) 1
   'Pearl of the Cape' 82

Clivia robusta B.G.Murray *et al.* **cover,** 7,
   13, **14,** 18, **21, 22,** 24, 26-28, 30, 32,
   35, 36, 40, 42, 45, **46,** 49, 50, 51, **54,**
   56, 103, 136, 137, 138, 139, 141, 144,
   149, 150, 154, 166
   'Maxima' 91
   var. *citrina* Z.H.Swanevelder *et al* **cover,**
   13, 21, **23,** 33, 39, 51, 88, 91, **92,** 98
   var. *robusta* 38, 51, **87,** 88, **89, 90,** 91, 92,
   172

Clivia Academic Committee, Changchun,
   China 17
Clivia Breeding Plantation, Japan 17, 123,
   126, 130, 131
Clivia Club see Cape Clivia Club
Clivia Cultivar Checklist and
   Registration process 103
Clivia Industrial Office, Changchun, China 17
Clivia postage stamp series **see** postage
   stamps
Clivia Society iii, 18, 186
Clivia Society, Changchun, China 17
companion plants 136
Condy, Gillian 10, 12, 18

Cowley, Welland 84
Crassula multicava **132**
Crinum 168
Crinum moorei 160
Cryptostephanus 24
   densiflorus 24
   vansonii 24, **25**
   haemanthoides 24
Curtis's Botanical Magazine 3, 5, 6, 7, 11,
   94
Cyprus Farm, Ofcolaco 16
Cyrtanthus 21, 24, 107, 168
Cyrtanthus elatus 160
   herrei 21, **41**
   obliquus 21, 41

D
damping-off fungi 165
Daruma hybrids 7, 29, 114, 117
Dracaena aletriformis 42, 45
   surculosa 123
Duncan, Graham **188**
dwarf clivias **see** novelty clivias
Dyer, R A 11

E
Edwards's Botanical Register 2, 3, 82
Erwinia carotovora bacterium 167
Erythrina caffra 88
Euphorbia lathyris 161

F
Fisher, Roger 77
Fitch, W H 6
Floral Magazine, The 4
Flore des Serres et Jardins de l'Europe
   15, 107
Flower Paintings of Katharine Saunders 8
flower stalk malformation 165
Flowering plants of South Africa, The 8,
   11, 72
Fusarium see damping-off fungi

G
gall midge fly 158
Garden, Major Robert Jones 7, 8, 94
Garden, The 8, 11, 64
Gardener's Chronicle, The 7, 58, 64
Gartenflora 5, 58
Gibson, Andrew 100

Giddy, Cynthia 66, 67
Gore, Pat 117
Green Girl see *Clivia* and *Hippeastrum* 'Green Girl'
greenfly see aphids

H
Haemanthus coccineus 11
Halleria lucida 48
Handbook on judging, showing and registration 19
Hart, M. 2, 82
Herbert, William 5, 84
Heritage Collection 121
Hippeastrum 16
Hirao, Shuichi 17, 114, 118, 119
honey bee pollination 42, 103
Hooker, William Jackson 3, 5, 6, 7, 82
Hosta 124
Howick Yellow see *Clivia miniata* var. *citrina* 'Mare's Yellow'
Hutchinson, John 1
hybrid clivias **cover**, 7, 11, 15-17, 19, 29, 46, 60-70, 80, 100-156

I
illegal plant collecting see plant collecting
Imantophyllum 3, 5, 8, 11
Imantophyllum? miniatum 58
Imatophyllum aitonii 3, 82
Imatophyllum cyrtanthiflorum 15, 107
insect pollination 42
intergeneric hybrids 16, 101, 103
interspecific hybrids 17, 19, 101, 106, 107
intraspecific hybrids 17, 19, 101, 112
IUCN categories 50, 51

J
Joubert, Elbe 9

K
Kew 3, 7, 8, 11, 119
Kirstenbosch iii, iv, 20, 51, 53, 66, 70, 79, 90, 91, 96, 110, 116, 117, 139, 149, 189
Knysna touraco 47
Koike, Toshio 114, 118

L
Lachenalia 40, 41
  bulbifera 41
  aloides **41**
leaf die-back see Agapanthus fungus
leaf miner 159
leaf spot 64, **164**, 165
lesser double-collared sunbirds 40
lily borer moth 60, 74, 80, 96, **157, 159, 160**
Lindley Herbarium, Cambridge 5
  John 3, 5, 7
locusts 160

M
maggots **160**
Major Garden's clivia see *Clivia gardenii* var. *gardenii*
malachite sunbird 40
Malan Charl 120
Mansell, Captain 7
Maputaland-Pondoland Centre of Endemism 94
Maytenus acuminata 48
McNeil, Adelaide 16
  Gordon 16
  Marguerite Rose 16
mealy bug 80, 158, **160**, 161
medicinal uses 49, 50
miracle clivia see *Clivia mirabilis*
mist clivia see *Clivia* x *nimbicola*
Miyake, Isamu 17
Miyoshi & Co. 119, 155
mole plant see *Euphorbia lathyris*
mole rats 161
Moon, H G 8, 64
muthi collectors 44, 50, 51
  see medicinal uses

N
Nakamura, Yoshikazu **16,** 17, 29, **109,** 110, 112, 119, 120, 122, 126
Natal drooping clivia see *Clivia gardenii* var. *gardenii*
natural hybrids 103, 104, **105**
Nectarinia amethystina 40
  chalybea 40
  famosa 40
  olivacea 40

nematodes **see** parasitic nematodes
Nerine 24
   bowdenii 160
Nieuwoudtville Bulb Project 52
night-flying moth pollination 42
Nimbicola **see** *Clivia* x *nimbicola*
Northumberland, Duchess of **see**
   Clive, Lady Charlotte
Duke of 3
novelty clivias 130, 131, 150
Noyce, Michael 70

O
O'Brien, James 11
Ogasawara, Ryo 17
Olea europaea subsp. *africana* 48
olive sunbird 40
Onie, Cameron 112
Orange List 51
Osborne, Melmoth 8

P
Papilio demodocus 41
parasitic nematodes 166
Phillips, E P 8
Phoenix reclinata 88
plant collecting 50, 51
Plant Life 16
Podocarpus elongatus 48
Pondoland Centre of Endemism 45, 88, 91, 94
postage stamps 10, **12**, 18, **19**
Powys Rogers, Mrs 7, 8, 64
Pretorius, Wessel 13
pygmies **see** novelty clivias
Pythium **see** damping-off fungi

R
Raes, Charles 15
Raffill, Charles 118
Red List 50, 51
red spider mite 161
Reed's Nursery, Wynberg 66
Rhizoctonia fungus **166**
rodents 161
Roemer, J A 5
Rourke, John P 13
Royal Botanic Gardens at Kew **see** Kew
Rumohra adiantiformis 42

rust fungi 60, 64, 166, **167**
Rutaceae 41

S
samango monkeys 47
Sasaki, Shigetaka 109, 112, 118
Saunders, Charles 8
   Katharine 8
   Rodney 66
   Wilson 11
scale 162
Scholtz, Clarke 42
Schultes, J H 5
Scott's Farm, Grahamstown 16
Sedum 124
slugs and snails 60, 162, 168
Smith Kenneth R. 103
   Claire Linder 14
Smithers, Peter 118
snout beetle 60, 80, 162, **163**
soft crown rot 60, 166, 167
Steedman, Andrew 5
Steenkamp, Kobus 121
stem clivia **see** *Clivia caulescens*
Strelitzia nicolai 44
sunbird pollination 40, 103, 133
swallowtail butterfly 41
Syon House 3, 82
Systema vegetabilium 5
Syzygium cordatum 88

T
thrips 163, 168
Thurston, Val 40
Tipperary Nursery 112
Transkei Yellow **see** *Clivia miniata* var. *citrina* 'Dwesa Yellow'
Traub, Hamilton P 16
Trichoplusia 42

V
Vallota 5
Vallota? miniata 58
Van Houtte, Louis 15
variegated leaves in *Clivia* **29, 36,** 108, 109, **123**-129, **134,** 135, 140, 141, 151-153, 155
vervet monkeys 47
Vico Morcote 11

virus 168
Von Regel, E A 5

W
Watson, William 7, 8, 11
whitefly 163
Winter, John **iii**, 51, 60, 110, 116
Wisley 145

USEFUL ADDRESSES

Membership of the Clivia Society will keep you in touch with other *Clivia* enthusiasts, whether they be keen amateurs, specialist collectors or professional researchers. The Society publishes a quarterly newsletter and a yearbook, holds meetings, arranges flower shows and supplies seeds and plants. More information on the Clivia Society can be obtained from their website: www.cliviasociety.org

Those wishing to join the Clivia Society who reside in South Africa are encouraged to contact the Clivia Society Secretary (see below). Those residing in Australia, New Zealand, The Netherlands, United Kingdom and United States of America should contact the relevant representatives in these countries, listed below.

The Secretary (Lena van der Merwe)
Clivia Society
P.O. Box 74868
Lynnwood Ridge 0040 South Africa
Tel/Fax: +27 (0)12 804 8892
E-mail: cliviasoc@mweb.co.za

Clivia Clubs and Interest Groups in South Africa

The Secretary (Joy Woodward)
Cape Clivia Club
P.O. Box 53219
Kenilworth 7745 South Africa
Tel/Fax: +27 (0) 21 671 7384
Cell: +27 072 487 7933
E-mail: capeclivia@ibox.co.za

The Chairperson (Andrè Calitz)
Eastern Province Clivia Club
Tel: +27 (0) 41 367 4476
E-mail: calitzap@absamail.co.za

The Chairperson (Hennie van der Mescht)
Free State Clivia Club
18 Mettam Street
Fichardt Park
Bloemfontein 9322 South Africa
Tel: +27 (0) 51 522 9530
E-mail: vandermescht@absamail.co.za

The Chairperson (Ida Esterhuizen)
Garden Route Clivia Club
P.O. Box 1706
George 6530 South Africa
Tel: +27 (0) 44 871 2214
E-mail: kobuse@xsinet.co.za

The Chairperson (Glynn Middlewick)
Johannesburg Clivia Club
2 Willow Road
Northcliff 2195 South Africa
Tel: +27 (0) 11 476 1463
E-mail: gcmidd@mweb.co.za

The Chairperson (Val Thurston)
KwaZulu-Natal Clivia Club
Tel: +27 (0) 31 763 5736
E-mail: thur001@iafrica.com

The Chairperson (Lena van der Merwe)
Northern Clivia Club
P.O. Box 74868
Lynnwood Ridge 0040 South Africa
Tel/Fax: +27 (0) 12 804 8892
E-mail: nclivia@mweb.co.za

The Chairperson (Louis Chadinha)
Northern Free State Clivia Club
P.O. Box 2204
Welkom 9460 South Africa
Tel: +27 (0) 57 357 6067
E-mail: lchadinha@xsinet.co.za

The Chairperson (An Jacobs)
Waterberg Clivia Club
Tel/Fax: +27 (0)14 717 3674
E-mail: johanan@esnet.co.za

The Chairperson (John Roderick)
Border Interest Group
P.O.Box 2429
Beacon Bay 5205
Tel: +27 082 567 7069
E-mail: jroderick@sainet.co.za

The Chairperson (Ian Radmore)
Lowveld Interest Group
P.O. Box 1146
White River 1240
Tel: +27 (0)13 751 2051
E-mail: ian@nelvet1.agric.za

The Chairperson (Joey Dovey)
Northern KwaZulu-Natal Interest Group
P.O. Box 8402
Newcastle 2940
Tel: +27 (0) 34 318 4179
E-mail: doveyw@telkomsa.netThe

Chairperson (Felicity Weeden)
Overberg Interest Group
P.O.Box 1468
Hermanus 7200
Tel: +27 084 5898 297
E-mail: fillylilly@lando.co.za

The Chairperson (Zanette Wessels)
Zoutpansberg Interest Group
P.O. Box 390
Louis Trichardt 0920
Tel: +27 (0) 83 326 6073
E-mail: pawrsa@mweb.co.za

International Clivia Representatives

Australia (Ken Smith)
593 Hawkesbury Road
Winmalee, New South Wales 2777
Tel: +61 2 475 43287
E-mail: cliviasmith@hotmail.com

New Zealand (Tony Barnes)
Ngamamaku, 1521 Surf Highway 45
R.D.4 New Plymouth 4061
Tel: +64 6 752 7873
E-mail: tony.john@xtra.co.nz

The Netherlands (Aart van Voorst)
Frederick Hendriklaan 49
Hillegom TE 2181
Tel: +31 252 529679
E-mail: a.v.voorst@freeler.nl

United Kingdom (Jaco Nel)
46 Atney Road, Putney
London SW15 2PS
Tel: +020 8789 2229
E-mail: neljaco@hotmail.com

United States of America and Canada
(Jim Shields)
P.O. Box 92
Westfield IN 46074
Tel: +317 896 3925
E-mail: jshields@indy.net

Clivia miniata Kirstenbosch Supreme strain, bred by the author at Kirstenbosch

Graham Duncan is a specialist horticulturist at Kirstenbosch National Botanical Garden where he curates the collection of indigenous South African bulbs, and the display inside the Kay Bergh Bulb House of the Botanical Society Conservatory.

His numerous popular and scientific articles on bulbs have appeared in leading local and international horticultural and botanical journals, and he is the author or co-author of six titles in the *Kirstenbosch Gardening Series*, including *Grow Clivias* that first appeared in 1999, was reprinted in 2002 and is produced here in a fully updated new edition. In 1989 he co-authored two major publications on indigenous bulbs, *Spring and Winter Flowering Bulbs of the Cape* with Barbara Jeppe (Oxford University Press), and *Bulbous Plants of Southern Africa* with Prof. Niel Du Plessis, illustrated by Elise Bodley (Tafelberg Publishers). He currently serves as an advisor to *Curtis's Botanical Magazine,* and is co-ordinator of the Kirstenbosch Gardening Series with Dr Neville Brown.

His special interest in the genus *Lachenalia* resulted in the publication of a popular guide to the genus in 1988 titled *The Lachenalia Handbook* in the *Annals of Kirstenbosch Botanic Gardens* and an MSc (*cum laude*) in Botany from the University of KwaZulu-Natal in 2005. He is currently working towards a monograph of the genus.

In 2001 Graham was honoured with the International Bulb Society's prestigious Herbert Medal.

For more information on fynbos plants visit

www.sanbi.org

click on

PlantZAfrica.com

FineBushPeople

fynbos seed suppliers
to South Africa and the world

- Seeds for fynbos garden enthusiasts
- Bulk seed suppliers to nurseries & farmers
- Uniquely packaged gift packs of fynbos seeds
- Books, fertilizer & primer for growing fynbos from seed
- Promotional and corporate gifts with personalized branding

Shop online at http://www.finebushpeople.co.za
or email julie@finebushpeople.co.za
fax: +27 21 701 3338

THE BOOKSHOP@ KIRSTENBOSCH

southern Africa's leading treasure house of natural history publications

Open seven days a week

10% discount to Botanical Society of SA members worldwide

tel **+27 21 762 1621**
fax **+27 21 762 0923**
email **kbranch@botanicalsociety.org.za**
www.botanicalsociety.org.za

Welland Cowley t/a Cape Flora

CLIVIA SPECIES AND HYBRIDS

Breeding of special cultivars from Welland Cowley:

- Large-umbel yellows with vico background
- Green-stem and other interspecific hybrids
- Peaches, Reds, Pastels, Pinks, Green throats!
- Development of *Clivia mirabilis* interspecifics
- *Clivia miniata* seed sold by the kilogram

We also supply seed of all varieties of *Strelitzia* including *Strelitzia juncea*!

See our Websites:

www.cliviasa.com

www.capeflora.com

Contact us on E-mail:

welland@cowley.co.za

Cell: **+27 82 511 8043**

P O Box 667, Sedgefield,
Southern Cape, RSA 6573

We are nuts about Fynbos!

COME TO US FOR THE BEST SELECTION OF . . . . .

Leucospermums
Leucadendrons
Proteas
Ericas

and many other Fynbos and indigenous plants

CALEDON FYNBOS KWEKERY/ NURSERY

Millstr 48, PO Box 624, Caledon, 7230 | Tel / Fax: +27(0) 28 214 1016 | Cell: 082 772 4681
E-mail: cfk@telkomsa.net   Website: overberginfo.com/fynbos-nursery/

CAPE SEED & BULB

Growers and breeders of special clivias and rare plants

Tel. 021 8852423
Fax 021 885 2421

P.O. Box 6363,
Stellenbosch, 7612
e mail: capeseed@iafrica.com
website: www.capeseedandbulb.com

Be wise. Fertilise all year round.

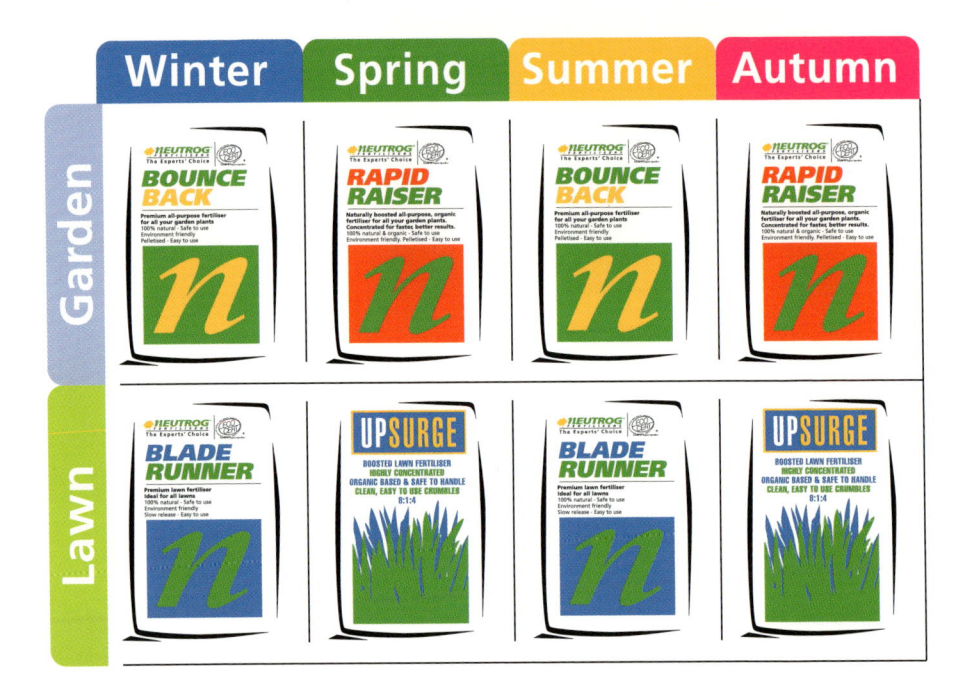

	Winter	Spring	Summer	Autumn
Garden	BOUNCE BACK	RAPID RAISER	BOUNCE BACK	RAPID RAISER
Lawn	BLADE RUNNER	UPSURGE	BLADE RUNNER	UPSURGE

You don't feed your family just once a year. So why starve your plants? Like all living things, plants need regular feeding for healthy growth. And Neutrog makes it easy for you to provide that balanced nutrition, with a year round fertilising programme. Just follow this chart, and see your garden grow as never before. Neutrog. At your favourite garden centre now.

Be wise. Fertilise all year round.

NEUTROG
FERTILISERS
The Experts' Choice

Neutrog Africa
PO Box 81 Philadelphia
7304
T **(021) 972 1958/9908**
F (021) 972 9909
E info@neutrog.co.za

www.neutrog.co.za

CLIVIA NOTES

CLIVIA NOTES